Steel Illusions

A cross between Mary Higgens Clark and Nora Roberts, R.Z. Crompton deftly weaves a suspenseful tale that will keep you turning the pages. Steel Illusions is a great second novel. I can't wait for more.
>Dennis O'Kane
>Book/Kids/Video Manager,
>Media Play

Excellent blend of fact and fantasy which captured my interest throughout.
>Tim Cochrane
>President,
>National Mountain Rescue Asso.

I enjoyed this fast paced thriller that involves the industries with which I've been involved my entire career. R.Z.'s second book piques the imagination. I'll look for additional books in the future.
>Michael Strauss
>President,
>Sand Springs Metal Processing Corp.

STEEL ILLUSIONS

To Helen,
I hope you enjoy this one as much as the first.
R Z Crompton
2-28-98

R. Z. CROMPTON

Zoller Publishing, Inc.
P. O. Box 461661
Aurora, CO 80046

Book cover design by:
Terry Bump Gilbert

This book is a work of fiction. Names, characters, and incidents are either products of the author's imagination, are used fictitiously, or are used with permission. Any resemblance to actual events or persons, living or dead, is entirely coincidental.

Copyright © 1996 by R. Z. Crompton

All rights reserved. Written permission must be secured from the publisher to use or reproduce any part of this book.

Library of Congress Cataloging-in-Publication Data

CIP 96-090520

ISBN 0-9649438-1-6

This book is dedicated to:

my wonderful husband who had a much better attitude on this book,

my daughters who helped bring the characters to life, and

my sister and partner.

"God may present the path of opportunity, but we must walk our own steps."

by R. Z. Crompton

Acknowledgements:
  Tim Cochrane, President
    Mountain Rescue Association
  Alex Majed, Manager
    Warren Duck Club, DoubleTree Hotel, Warren Place
  Caché
  Wnyfrey Hotel, Birmingham
  Steve Jones Stables, Beaver Creek, Colorado
  Lakota River Guides, Vail, Colorado
  Montaneros, Vail, Colorado
  Tommyknockers, Idaho Springs, Colorado
  Timber Ridge Snowmobiles, Avon, Colorado
  Christi Sport, Avon, Colorado

Written Resources
  The Avalanche Book, Armstrong and Williams
  AMC Guide to Winter Camping, Gorman
  Birmingham and Jefferson County, Atkins

Steel Illusions

CHAPTER ONE

"Pete, what's going on?"

A handsome face looked up from the papers on the table in front of him. His concentration was focused on the numbers, not on his wife. "What are you talking about?" He asked the question without waiting for the answer before tilting his head down to look once again at the papers in front of him.

"Pete," she said with more emphasis in order to capture his undivided attention. "You've been in the twilight zone all evening. I'm not a fool; we've been married long enough for me to know when you're upset. Don't you want this baby?"

That question grabbed his attention, and he looked up. "Of course, I do. I can't believe you'd even ask such a question. Really, there's nothing wrong." Pete purposely tried not to meet his wife's gaze; he was afraid she'd be able to tell he was being less than honest. So he again let his eyes drop to the papers in front of him.

Amy watched the tall muscular man squirm on his chair. "Peter, you always fidget when you're trying to hide something from me, and I'd understand about not wanting the baby. I've had my own doubts."

"Really, Honey. Sitting here agonizing over each of these bills always makes me an old grouch."

"I know, I know." Amy's voice trailed off as she turned around allowing her eyes to gaze out the kitchen window into the darkness. Fear seemed to be tightening its grip on her emotions, and she didn't know why. Amy didn't turn from the window; she was certain he wasn't telling the truth. All night he had been pacing like a caged animal desperate to find the open door. She knew he was up to something, and it wasn't good.

"C'mon, Amy. You know Molly's hospital bills are killing us." Wanting to comfort his wife, Pete walked over to put his arms around the petite, blond woman standing in front of the kitchen sink. He bent

Steel Illusions

down to kiss her tenderly on the top of her head. The soft curls tickled him, and he had to use one hand to scratch the tip of his nose while the other reached down to pat his wife's still flat belly. He loved her and the small life growing inside of her just like he had loved Molly. He knew he'd give his own life to save Amy or one of his children. He also knew he'd do anything, anything at all, to provide a good life for his family. *I will take care of you; one way or another, I will take care of you.* Pete set the covenant with his wife without saying a word.

"When Molly died, there was a black hole left in my soul that's never healed. It's only gotten deeper. I'm just preoccupied with my own thoughts tonight. This baby you're carrying has given me something to hold on to: something to look forward to. The past has haunted us long enough. I refuse to bring a child into this world with the black cloud of debt hanging over us. If we don't do something now, our baby will always live with the burden and deprivation caused by that damn monster which destroyed our little girl. I don't want you to live in this hole of an apartment for the rest of your life, and I certainly don't want my children to grow up here. I hate this place."

Pete was pacing again, cracking his knuckles the way he always did when he worried about money. "We work day after day just to barely cover the bills. Where's our future? Here?" He was nearly yelling as he swung his arms out in front of him. "This horrible little apartment with its chipping paint, leaking pipes, and man-eating cockroaches? No Thanks!"

Amy walked up to her husband and cuddled into the strong arms as she whispered up to him, "I know. I know." The words vibrated through her mind like the deep hollow tones of a base drum. It seemed she had repeated *I know* a thousand times since Molly had died. Every time Pete started complaining, she simply tried to comfort him by agreeing.

R. Z. Crompton

Trying to draw new strength from the warmth of her husband's body, Amy leaned in closer to his firm, muscular chest. She understood her husband's desire to make a new life for them, but she couldn't help shake the ominous feelings which had been with her since he'd gotten home. Maybe her negative feelings were due to the fact that Pete wouldn't tell her what his plan was; and then again, maybe her hormones were playing with her emotions. She'd debated day after day with herself about whether or not to even tell him about the baby. Some days she had hoped for a miscarriage. Abortion had crossed her mind more than just once, but she wanted the baby so badly. After Pete had caught her throwing-up more than once, she couldn't hide the fact she was pregnant anymore. Now his mood swings were nearly as drastic as hers had been. Amy rationalized that her husband deserved his own time to deal with the reality of what a baby would mean to them financially and emotionally.

"Now, you go to bed; I'll be in after I get some work done. Tomorrow's a working day." Pete gave Amy a tender kiss and sent her off to the bedroom before returning to the pile of bills setting on the small table in the corner of the room. He hated the tiny, dingy kitchen where they'd been eating their meals for the last two years. Every time he drove his pick-up into the parking lot, he cringed at the sight of the run-down complex. They had worked so hard to build a life: a good life. Before Molly's illness, Pete had everything he'd ever wanted: a beautiful wife, a wonderful child, and a good job which left him feeling exhausted but satisfied at the end of the day. Then Satan, in the guise of an invisible, savage mouth, devoured his daughter.

After the long battle with leukemia, Pete had lost his tiny angel to the ferocious jaws which had masticated her body little by little, day by day, until the tiny frail bones nearly crumbled under his touch. In fact, by the time God had finally released Molly from her suffering, Pete was glad the torture was over. He'd never forget the helplessness

Steel Illusions

he'd felt every time a nurse or doctor stuck her little body with another needle and her large brown eyes looked up at her daddy waiting for him to do something that would save her from the pain. Tears flowed endlessly as he and Amy had watched the hospital staff poke and prod their only child. Pete knew now, that at a certain point, he should have just taken his darling baby home to let her die in the peace and love of her mother's arms; but he had continued hoping, until the very end, that the doctors would come up with some miracle to save her; some miracle which would have stopped the slow consumption of her body.

Unfortunately, the appetite of the invisible invader hadn't been satisfied with devouring the innocent body of his daughter, it continued to wreak financial havoc on him and Amy. Some evenings Pete just sat and wondered about how the illness had reached so far into their lives.

After they had buried their darling little girl, Pete nearly declared bankruptcy, but his pride kept him from giving in to the continuing torture of the disease. Insurance had paid the majority of the hospital and doctor bills; but there was still an enormous debt for them to cover. Debt was all relative like most things in life; and that probably frustrated Pete as much as anything. Some other guy with a high paying job would have covered the bills instantly or, at least, in a matter of months or years. But Pete wouldn't cower in front of the enemy, and that's how he saw the pile of bills. Instead, they sold their new house and car, and moved into the low cost housing district across town. The financial quagmire, however, kept pulling him in deeper and deeper. He was looking at a lifetime of hospital bills; and each time he sat down at the little table in the dingy kitchen to face his enemy, the pain of his loss was relived with every check he wrote.

When Amy had told him she was expecting again, Pete didn't know if he should be happy or sad. The thought of bringing another baby into their life brought so many doubts. He felt desperate to lift the shadow of debt before this new child was delivered to him, and suddenly it seemed the opportunity was right in front of him.

The superintendent had put out a night letter containing the information about the test. Some special scrap was going to be delivered to the melt shop the next afternoon. His instructions were easy enough. All he had to do was make sure the majority of the special delivery went into one heat rather than being mixed in with the scrap on the ground. The instructions referred to a special alloy grade heat: some kind of high-chrome shit Pete assumed. Some of the scrap might appear to be closed containers, but everything would be checked before it arrived. There was nothing unusual about the night letter, except that it usually had more to do with the midnight shift than the next day shift. Special grades of steel were made on a regular basis. Closed containers were the danger zone, but there was no reason to question Bradford. Kevin Bradford, the meltshop superintendent, was a straight shooter and fair with the crews. The guys respected his leadership and his steelmaking ability.

It was the possibility of having closed containers which gave Pete his idea. He knew it was kind of dangerous, but steelmaking in general was dangerous. Small explosions were not so unusual. Lots of the guys had been burnt from time to time. If Pete could get the details lined up just right, the amount of money he could collect for the day's work would be enough to pay off the bills and give them a new start. He would finally be able to think back on the good times he'd had with Molly without the anger which always followed when reality bit back at him.

Pete paced the kitchen again. It was easier for him to think through the details of his plan with Amy tucked safely in bed. She'd be angry if she knew what he was planning. Pete regretted keeping his decision a secret from her, but she'd do everything she could to talk him out of it. She didn't understand the frustration he felt every time he paid another hospital bill. He was barely covering the interest. Pete sat down at the table to write his wife a letter of instruction just in case his plan went awry. He'd mail the letter on his way to work tomorrow. If nothing happened, then there was no need for her to know about anything. If things went terribly wrong, she had a right to know what he had been trying to accomplish. When the letter was finished, Pete turned off the light and went to join his wife in bed.

****

In a luxury high rise apartment in downtown Tulsa, a tall, swarthy man stared at his reflection in the mirror as he slicked the sides of his coal black hair behind his ears. "That son-of-a-bitch starts to suffer tomorrow," he said aloud. His eyes opened wide with anticipated pleasure.

"You're such an asshole," the reflection answered back at him.

"I know. Isn't it fun." The conversation proceeded between the man and his reflection.

"Pay backs are hell unless you're the one dishing it out," the alter ego replied to the man standing in front of the mirror.

"He's given me enough grief to last forever. One of us is going to be fired, and it's not going to be me. I want to watch him dangle on a hook like a worm being hacked off bit by bit." The mental image played in his mind causing a smile to slither across his face.

"I hope Jean is ready for a celebration. I want a chilled bottle of Dom Pérignon and a good roll in the sack." Moving away from the mirror, the man headed for the door turning off the light as he left.

****

R. Z. Crompton

Kevin Bradford felt the extra surge of vibration under his feet; and even before the sound was audible, the hair stood up on the back of his neck. Years of firsthand experience told Kevin this explosion would be heard all the way to town. The fire department would probably be on its way before he could even have someone call for help. Kevin wasn't quite out of his chair when the explosive sound waves vibrated through the area; it was deadly.

Kevin was headed out the office door as the thrust of smoke and ash hit him in the face. He visualized eighty-five tons of molten steel flying through the air. Crossing the east end of the charging floor in less than a dozen strides, he entered the furnace area just as a secondary explosion ripped through the slag door. Intense heat singed his eyebrows and the hair on his arms while he tried to assess the situation. He covered his mouth and nose to keep from breathing in the toxic fumes. Liquid steel and slag were being thrown through the air only to rain down on every man within range. Some of the men crouched down, trying to protect their body with their flame retardant coat, while others ran trying to escape the flying fire and piercing shrapnel. Immediately after the echo of the explosion subsided, a haunting silence hung heavy in the air.

As Kevin entered into the darkness, screams of pain chased the silence away. Months of mill dust, the byproduct of steelmaking which collects in the rafters, was sifting down making it almost impossible to see anything but the glow of red. Men whose clothing was on fire dropped to the ground to roll out the flames. The unmistakable chemical stench released from the flame retardant gear gave Kevin some relief because he knew the special coats were protecting human flesh. The next odor Kevin recognized was that of burning hair, probably his own.

Kevin started yelling commands at the men who appeared to be okay.

Steel Illusions

"Joe, call 911, get EMSA and the fire department out here." Kevin hated the fact that the mill wasn't big enough to have its own fire crew on site.

"Bob, take two or three guys and get the fire extinguishers going. C'mon, Dan. Help me get Pete calmed down and comfortable." The two men rushed over to the injured man on the ground. Kevin had seen Pete covered in flames running from the furnace.

"Jesus, Dan, why didn't he have his mill coat on?"

"Don't know, Boss."

"Why was he on the furnace floor instead of in the pulpit?"

"Don't know."

"I can't see - can't see!" Pete shrieked at Kevin.

"I know. Try to be still. We'll take care of you." The synthetic material of what had been a shirt had melted into Pete's skin.

Dan could tell by the way Kevin was inspecting Pete's body what his next comment was going to be. "He knew better, Boss. You always said 'wear cotton.' And his coat? I don't know. He knew better! Mother of God, he knew better," the man's voice trailed off sadly.

"Then why in the hell doesn't he have it on. Get the fire blankets out of the first aid locker. Hurry!" Dan was glad to run for the locker that was kept in a protected area behind the pulpit. It'd taken all of his control to keep his stomach down while he squatted next to Pete. The smell of burnt flesh permeated the entire shop. There wasn't a man in the place who wasn't familiar with the pungent odor and the guilt it sparked because each felt relieved it wasn't his own flesh which had been melted away by the molten steel.

"Get the saline solution. Hurry up! Some of the guys will need to wash out their eyes so they won't burn. The dust falling out of the rafters can irritate the eyes and lungs, too."

R. Z. Crompton

"Here, Boss. Wash Pete's eyes first." Dan handed a bottle of saline solution to Kevin along with some cool packs that would help, marginally, to give Pete some comfort. They covered him with the special blankets to keep him from going into shock.

Pete was still conscious, and they didn't want him to know how badly he was hurt. Maybe it was a blessing the fire had blinded him. Otherwise he'd be able to see how seriously his body was burnt. The pain is bad enough, but somehow seeing the destruction of the flesh makes it worse. All of the hair on his head was gone and so was most of the skin. Red, exposed muscle tissue was mixed with the black char of dead skin. The foul stench of death hung in the air, and every man recognized it.

The excruciating pain caused constant moaning to gurgle from Pete's mouth. Kevin was sure there wasn't a more painful way to die. He'd seen enough third degree burns over the years to know, with reasonable certainty, that Pete didn't have a chance in hell. No matter how quickly Pete was taken to the burn unit at Hillcrest Hospital, the boy would suffer a slow agonizing death as his major body organs gradually ceased to function. Meanwhile, the pain of the external burns would eat away at his sanity. As the burnt flesh tightened up, breathing would become nearly impossible. His lungs would slowly fill with fluid which, in turn, caused the coughing that ripped at the charred tissue. The kid would know in a few hours that it was just a matter of time before he would die. If the doctors didn't tell him, his body would.

"Damn it!" Kevin whispered to himself. "Why didn't you have your coat on, Kid? Why?"

Kevin looked around at what was left of his melt shop. The building stretched nearly the length of two football fields. Fires were still burning everywhere the molten steel and slag had fallen. Glass had been shattered immediately and launched like tiny missiles through

Steel Illusions

the air causing as many injuries as the flying steel. Explosions and chemical burns were nothing unusual in a steel shop, so it was a matter of necessity to put the men's safety before production and quality.

"Bob, how are the guys?"

"Most of them are shaken up, but okay. Only a couple will need to go to the burn unit with Pete. We were lucky everybody was wearin' his mill coat. Everyone except Pete."

"I want everyone checked by the paramedics when they get here. If there's any doubt, then off to the hospital he goes. Make sure of it, Bob."

"You got it, Boss. Anything else?"

"Get some of the guys busy with those fire extinguishers."

"Right away." Bob had already directed the uninjured to start putting out the fires. If they didn't get things under control, there'd be more explosions. Most of the damage to the shop and equipment had been done at the moment of the first explosion. However, any mobile equipment which had caught fire could still cause massive damage or injury if it exploded.

The ambulances had arrived before the fire trucks did. Pete was made as comfortable as possible before he was loaded into the back of the vehicle. Kevin sent a half a dozen others along to the burn unit to be taken care of. Several others were treated for minor cuts, scrapes and burns by the local medic on hand at the mill clinic. Johnny was sent off to another hospital for his broken leg. He had fallen from the stairs outside the pulpit trying to get away from the flying debris. The fall probably saved his life because the furnace control pulpit had been completely destroyed by the blast.

"Take care, Johnny. We'll check on you in a couple of hours." Kevin took the man's hand before he was loaded into the second ambulance.

"Boss, will you have somebody give Cindy a call. I'd like to have her meet me."

"Sure thing, Kid. I'll call her myself."

"Thanks, but you better call Amy first. I saw Pete running from the furnace. She'll be lucky if she can get there before he..... ya know."

"Yeah, Johnny, I know." Kevin tried not to look at the bone protruding from the man's leg. He was going to hurt like hell shortly.

Kevin didn't know where to start. The fire trucks were finally pulling through the gates. They'd make sure all of the fires were put out. "Bob, get some of the guys to move out any equipment they can before everything is doused with water and foam. We gotta save what we can. I'm gonna make some phone calls. Yell if you need me."

"You got it, Boss."

Kevin had the job of making the phone calls to the families of the injured men, but it'd been a long time since he'd made a call which weighed as heavy on him as the one he was about to make to Amy. Part of what happened to Pete was his own fault, and that would be hard for her to accept. Her natural reaction would be to blame anyone or anything other than her husband. Over the years, Kevin had learned there was always an explanation for an accident, even if it wasn't the explanation people wanted to hear. "Shit" didn't just happen like some of the guys claimed; some action, good or bad, was the root cause. Kevin was sure he'd find out the truth if he asked enough questions. He knew there had to be a reason why that kid was down there by the furnace without his flame retardant coat on. There had to be a reason; and that reason would cost Pete his life.

CHAPTER TWO

"Mom, I can't believe it's so beautiful today, and I almost missed it."

"I know, Meg. It's a good thing you got that last flight. Ya know, after fighting the constant snow flurries the last two days, the sun is going to feel sooooo good. It was certainly worth waiting for. See the hot air balloons taking off?"

"Yeah, I love watching them take off in the early morning. They're so colorful. I don't remember that water being there?" Megan pointed to a small glassy reflection a short distance from the condo.

"Nottingham Lake. It's hard to see unless the sun is shining just right. It's more of a pond than a lake. You and your sister used to play in the park next to it before the hotel was built."

"I don't remember. Does the name have some special reference to Robinhood and his merry band of men?"

Laura chuckled at her daughter's reference to the legendary Prince of Thieves. "No, Silly. It's named after the Nottingham family which settled here in the early 1900's. I think Alan Nottingham and some of his family still live around here. Years ago, he was one of the owners of Cassidy's Hole in the Wall."

"I like that place. The food is great and the building looks so old."

"The restaurant hasn't been around as long as you think."

"With a name like Cassidy's Hole in the Wall and the rustic ambiance of the bar, it has to be at least fifty or sixty years old."

"Sorry, Dear. It's only been serving the public for about twenty years."

"No. Really? How do you know all of this stuff?"

Steel Illusions

"We've been coming out here for a long time. I asked a few questions and read the papers. It's amazing what one can learn from the menu besides what's being served for dinner. Now, hurry-up. The slopes will open soon, and the fresh powder will be perfect today." After Megan turned away from her mother, Laura Bradford lingered at the patio door. She loved the view of the Rockies from this balcony. Every winter for the last ten years or so, she and her daughters had spent spring break skiing at Beaver Creek, Colorado. They had tried some of the other popular places in the area, but Beaver Creek always drew them back.

Megan, Laura's youngest daughter, had flown into the Denver airport from Texas A & M. She hadn't been sure she'd be able to leave until she'd gotten on the plane. As a first year vet student, she had to take her limited observation whenever she could get it. Dr. Alice Blue was her mentor and helped Megan get in a little extra time in surgery when most of the other students had gone home for a holiday. A last minute surgery on Hattie, the ostrich, had almost kept Megan from this year's fun.

Every year it was getting harder for the three of them to get away; but once again tradition prevailed. A successful operation, dash to the airport, several hours on stand-by, and Megan was able to join her mother and sister. This year a couple of extra bodies had come along for the fun. Krista, Laura's oldest daughter had invited her best friend and her boyfriend to join them. Laura enjoyed the temporary addition to the family. She missed Kevin, her husband of more than twenty years, when they were gone; but he didn't like skiing. This was a trip only the girls made, and they always enjoyed themselves. However, this year's trip had a slightly different twist to it because, for the first time, a man was along. Poor Brad had his hands full putting up with four women. He was easy going and good natured; and, as a part-time family member for the last few years, he gallantly took all the female harassment and frequent shopping trips in stride.

"Brad, get out of the bathroom. Your time's up!" Krista leaned on the pale, blue wall next to the bathroom door.

"Tough luck. You were the last one up; you're the last one in." Brad sent back his quick response.

"Please, Brad. Mom's almost ready to go."

"Then use her bathroom."

"I can't. Chelli's in there."

"Fine! You can share with me or with her. Make your choice. The door isn't locked."

"Mother, why does he get to use the bath as long as he wants; and the rest of us have to share?"

"Well, let's think about it, Dear. He's the only male in the condo, and there's four females. You were the last one up, so you get to share. I agree with Brad. After all, you invited him."

"Fine! Fine! Chelli, hurry up. It's my turn."

Laura smiled as she watched her daughter march down the hallway toward the master bathroom. She was sure her long-legged girl would be up at the crack of dawn tomorrow. Krista didn't sleep in often, and today she didn't like being last in line and out of hot water.

Laura poured herself another cup of coffee. It'd be at least forty-five minutes before the three girls had finished their required time in the bathroom and were ready to go. Besides, she wasn't in any hurry. The chair lifts wouldn't start running for another hour; and in spite of, the beautiful blue sky and sparkling snow, it'd still be bitterly cold until the sun was higher in the sky.

"Mom, are we packing a lunch today?" Megan asked as she walked into the living room. Her beautiful, light brown hair curled slightly around her face and then cascaded down her back. Even though yesterday's sky had been cloudy, there was a hint of sun on her rosy cheeks, or maybe it was just the excitement of another dazzling day in the Rockies. Whatever the cause, Laura marveled at the natural beauty of her daughter.

"Let's just take some snacks, Meg. We'll have lunch at the Spruce Saddle. I'm sure you're as tired of sandwiches as I am. By lunch time, it's supposed to be nearly forty degrees; eating on top of the mountain sounds like more fun."

"Terrific! I love eating up there. By the way, did you pack my black sweater?"

"I don't remember, Dear. We were in such a hurry to pack warm clothes for you that we just threw some warm looking stuff in a bag. Wear whatever you can find."

"Mom, 'whatever I could find' might have worked on the slopes when I was ten, but at twenty-two, dressing is a little more important. Just ask Krista. She takes forever to get ready for skiing, and she already has a boyfriend."

Just as Megan finished her comment, Brad walked around the corner of the room. "Yes, she does take forever, but she knows I appreciate the way she looks. Besides, I'm certain she wants to make sure my eyes don't wander too far," he finished his comments with a sly grin and a sparkle in his eyes.

"Brad, I can't believe you said that!" Krista yelled from the bedroom. "Maybe, I'm checking out my other options, too. After all, we're dating, not dead!"

"Touché, my Dear," Brad yelled back at her. Then he turned to Laura and added softly, "Ya know, the 'walking' scenery here really is fantastic."

"I know. I'm not dead either. It's not the looking and appreciating that's dangerous. It's the touching."

"I agree. One hundred percent!"

Krista hadn't missed as much of the conversation as Brad had wished, and she added her feisty comment as she entered the room. "So do I. I also like being considered part of that scenery you, and I assume, others are enjoying," she finished by tossing her long auburn curls over her shoulder. Carefully applied make-up high-lighted the cheeks and toned down the traditional goggle marks left from skiing the day before. Her beautiful blue eyes were brilliantly accented by the blue ski sweater she was wearing. She looked like she had just walked off a page of the latest ski magazine.

Chelli, Krista's best friend, made her first appearance of the day and added a quick comment as she opened a can of Coke. "Are you guys going to waste the day talking about the scenery, or are we going to join the fun?"

"Good morning, Michelle."

"Wow, Chelli! There's scenery and then there's major land marks. Why are you looking so fine today?" Brad asked the dazzling young woman.

"I just happen to be planning on enjoying my own little section of male beauty today. I have a date for lunch."

Everyone was a little surprised that she hadn't said anything the night before, and Chelli could tell by the raised eyebrows that questions were running amuck in all of their heads. So before any one could ask, she added a minor piece of information. "He just doesn't know it yet."

Laughter filled the room until tears were threatening to destroy all of the carefully applied make-up. Krista was the first to gain enough control to add her comment. "Only you, Chelli, can pick a guy and be dating him before he even knows who you are."

Everyone nodded in agreement.

"I resent that. He smiled at me yesterday, and we shared a brief conversation."

Steel Illusions

"What, exactly, did he say?"

"Well, it was something like 'Hurry-up or you're going to miss the chair.'"

Now it was Brad's turn to add a sarcastic remark, "Definitely true love, Chel."

"He's really cute and has dreamy brown eyes. I know where he's working today; and one way or another, he's having lunch with me."

"He's working?" Megan questioned.

"Yeah, he's a ski instructor."

"Wait, let me guess. You're taking a lesson? The only real question is whether or not you're going to be on the bunny slope with a group of kids or rolling head first down the moguls?"

"Really, Brad, you never did give me enough credit. I made sure I signed up for a private lesson and specifically requested the teacher I preferred."

"My money's on you, kid," Laura smiled at the attractive young woman standing in front of her. Sun-kissed cheeks added to the beauty of her green eyes. Her five and a half foot, slender frame was the perfect host to the body fitting ski fashions. This young lady could definitely turn a man's head, Laura thought as she walked to the refrigerator.

"Now, how about something more nutritious than a Coke for breakfast. You won't have the energy to find the guy if you don't eat."

Krista helped Laura set out some fresh fruit, bagels with cream cheese and juice for the five of them. Everyone had a bite to eat and helped clean off the counter. The slopes would be covered with fresh powder, and only the early birds would get the full pleasure of the early morning snow.

"Don't forget the sun screen. We'll need it today. Are you finally ready, Chel?"

Laura opened the door and headed out into the cool crisp air.

Brad held the door so Krista could walk past him. "You know I love you, don't you."

Krista smiled back into the handsome brown eyes, "Of course, I do. If I didn't, you wouldn't be here."

"By the way, I don't mind if you look at the scenery, just remember 'the last dance is for me.'"

Krista smiled at the reference to the old song and added, "Exactly. I don't care where you get your appetite, just make sure you come home to eat."

"Really, Krista, where do you come up with such lines?"

"I know you appreciate my quick wit. Daddy taught me good. Didn't he?"

"No, he taught you 'well.'"

"Brad, my mother taught me 'well'. Daddy teaches 'good.'"

"You're right about that. When I think about your folks, your dad does have a slightly slanted use of the English language."

Krista leaned over to receive the kiss Brad intended to plant on her lips just as Megan turned around. "Come on, you two. Can't you do that some other time? We're on a mission here."

"Krista, we really need a little privacy. Let's not go today. I love your mom and sister and Chel too, but three days with absolutely no time alone is more than a man can take."

"We have to go. Today may be the only day we get any sun. Besides, do you realize how much crap we'd get from those three if we didn't go? How about skiing back here for lunch while the others eat on the mountain?"

"That'll work. They'll never miss us. It's a date." Brad's smile couldn't be missed as he walked up to join the others.

Laura suspected what was going on and kept quiet, but Megan couldn't resist. "What are you two planning? Brad, you look guilty as a two-year old caught with the crayon in his hand and the new art work on the wall behind him."

"I'm not guilty. Just looking forward to a wonderful day." The smile on his face didn't waver as he tightened his grip around Krista's waist.

****

"Mom, look at all the snow across the peaks. I've never seen so much." Megan was sitting in the four-man lift next to Laura on their fourth trip to the top. Krista and Brad were on the other side of her. Chelli had taken off to find her instructor and lunch date as soon as she'd paid for her lift ticket.

Laura looked across the sparkling valley and up the opposite range. She'd never seen snow piled so high against the trees. What had seemed like normal towering pines yesterday were only black specs on the opposite range. It was difficult to get a good perspective on size and shapes because of the distance involved; but the snow appeared to be rolling glistening waves, like the surf of the ocean. The frozen waves rose and fell in patterned succession across the top of the peaks. Jagged harsh rock of two days ago was now a smooth sea of white.

Suddenly a cloud of white powder erupted from the top of the mountain destroying the pristine view. Laura quickly pointed out the spectacle to the others. Silent waves of snow which had innocently covered the opposite range were now billowing down the slope, ripping out trees and dragging large boulders on its trek to the valley. It was a magnificent display of power which left each of them speechless as they watched the destruction scar the range. As the snow raced down the slope collecting debris, the roar of its power echoed across the Rockies. Several seconds passed before the echoing noise subsided, leaving the four witnesses shaken.

R. Z. Crompton

"Holy Shit!" Brad was the first to offer his impression of the episode.

"No kidding," Krista added after she realized she'd been holding her breath. "Mom, if anybody was down there, they're dead now."

"Don't worry, guys. I'm sure the rangers are in control. Ya know, they often set off special explosives after a big snow just so they can control the danger of a possible avalanche, especially where there's people around." Laura tried to ease their concern, but she was only assuming the rangers were in control.

"That doesn't mean it's not possible."

"Megan, don't be such a 'fraidy cat'. We ski at the top of these slopes. We could see if snow was hanging over the edge of the ridge. Besides, the rangers are going to make sure things are safe here, that's their job. Isn't it? They probably set off the one we just saw." Brad tried to put her fears and his to rest. He didn't want to ski all day looking over his shoulder.

"I know that, but I also heard about a bunch of people killed in the Alps last season when an avalanche surprised them. Those people didn't even know there was any danger. There was no snow hanging precariously over the top of a peak. It just fell down the side of the mountain and buried them. So don't make fun of my fear." Megan haughtily tilted her chin up meeting Brad's sarcastic eyes.

"How do you know so much about an avalanche in the Alps?" Brad's patronizing tone further irritated Megan.

"We had some visitors from southern France at school last month, and they told me about the killer slide. Most of those people were acting just like you are: No Fear! Now, over twenty of them are dead. Some of the bodies weren't found until weeks later."

Laura wanted to calm her daughter's fears. "You're right, Dear. An avalanche is nothing to take lightly; however, I really don't think you need to worry here. If there was any danger, I'm sure the slopes would be closed. Now try to relax and have some fun. Last one to the lodge buys lunch."

Everyone prepared their skies for unloading as the chair lift approached the magic spot, but the scene of the Mother Nature's power played over in their minds. Megan, nearly losing her balance as she stood up from the chair, was still visibly shaken by the episode.

"Meg, you okay?" Laura questioned.

"Yeah, I just need a minute. Where are Brad and Krista headed?"

"They're having lunch back at the condo. We'll meet them later this afternoon at the fire pit."

"I see. There wasn't enough action on the slopes, right?"

"I think there's too much action. Too many people for their taste. They, ahh...want some time alone. After all, we've been with them every minute since we got here. C'mon I'm hungry. Let's go have some lunch."

Gravity was their partner as the two women waltzed down the slope, meandering across the side of the mountain enjoying the sun, the deep blue sky, and crisp white snow. However, the mouth watering aroma of the grilling burgers subconsciously caused Laura to quicken her pace. Guiding herself to the center of the ridge, Laura knew she'd find the lodge marking the midpoint of the slope. It was a favorite lunch spot with the skiers especially when the sun was shining. Laura and Megan propped their skis against the rack with the dozens of others coming down to whet their palate with the tantalizing burgers.

"Mom, I didn't realize how hungry I was until I got a whiff of that aroma. I feel like I've been skiing for hours."

"I know what you mean. Ya know what's funny though? The actual taste of the food is never quite as good as the anticipation it arouses while it sizzles on the grill. Water or juice to drink?"

R. Z. Crompton

"I'll take water. Please grab a chocolate chip cookie for me. I'll get the burgers since my anticipation has been aroused beyond my taste buds capacity to appreciate," Megan teased.

As Laura approached the table where her daughter was sitting, she was surprised to see her with a well tanned young man. There was no mistaking the coat: he was part of the ski patrol staff for the resort. Laura assumed Megan's avalanche anxiety was behind her; and, just like Chelli, she was focusing on her "mission" of the day. However, the topic of conversation surprised her as she came within hearing distance.

"Really, Miss, you don't need to worry about avalanches on this side of the mountain. Even though we're on the leeward side, any possibility of an avalanche is monitored daily."

"Wait a minute, 'even though' insinuates that leeward is dangerous. I may not know much about the mountains, but I do know that the lee side of a ship means the safe side. Doesn't that mean the leeward side of the mountain should be the safe side?"

"The lee side is considered the safe side if you're only talking about the wind. The harsh mountain winds blow against the windward side of the mountain, generally the west side, stripping off anything it can. It picks the snow up off the windward side, and guess where those innocent little flakes are dropped."

"On the leeward side?"

"You got it. What starts out as a six inch snow fall on the west side of the mountain can become a foot and a half or more in drifts on the opposite side. The irony is that the safe side of the mountain, where the snow settles, makes the leeward side the most prone to avalanches. The gradual slopes and deep bowls make this area a favorite for the skiers, so we mark off the best trails and watch for danger spots."

Steel Illusions

"What about the windward side? There was lots of snow over there; we were skiing the back bowls at Vail yesterday."

"Well, when there's a hundred inch base, it looks like there's a lot of snow everywhere."

"I guess so. But I don't understand why some places are so much more dangerous than others."

As the extremely handsome man pointed to the ridge where they had seen the avalanche, Laura interrupted their conversation. "Still worried, Meg?" Laura asked sitting down next to her daughter.

"Hi, Mom, this is Luke. He's part of the ski patrol. Luke, this is my mother."

"Hello, Luke. I figured out the ski patrol part. The coat's a dead give away."

"Pleased to meet you, Ma'am. I was just trying to put your daughter's mind at ease."

"I don't understand why some places are more dangerous than others." Megan seemed obsessed with the topic. Laura listened intently to Luke's oration about the snow. She could tell he was as serious about the topic as her daughter was, and she was especially glad the man didn't take her daughter's fear lightly.

"The wind is like a master architect. It builds the snow drifts layer by layer using new angles and dimensions constantly. The powerful gusts scoop up the freshly deposited flakes and drop them into the gullies on the downward slopes. The wind can turn a relatively light snowfall into deadly drifts on the opposite of the mountain. When the slope is nearly vertical, it's easier for the snow to build up a heavy top load. Now if the wind has been blowing really hard, a ridge of snow can form which actually leans over the top of the mountain. It's kinda like the curl in a wave. We call it a cornice. That's the most dangerous type of drift."

"Does a person caught in an avalanche always die?"

"Of course not. Lots of people get caught in slides especially little ones called sluffs. I was caught in a slide last year."

"Really?"

"I don't want to mislead you; I was terrified. The speed that I traveled down the face of the mountain was incredible. It took only seconds for the snow to tumble me about five hundred yards."

"Why weren't you killed?"

"Most of us who work up here take the local training about survival in the back country and it includes living through an avalanche."

"So it could happen to any of us, but the basic skier would die."

"It's not that drastic. If you stay on the marked trails, you really don't have to worry. It's the 'macho' type, skiing in the unmarked areas, who's in danger. A couple of us had gone after three teenagers who just wouldn't follow the rules. I remember thinking, while I was rolling down the slope, that I wanted to ring the neck of the kid I was chasing, but I didn't get the chance. The mountain gave a much harsher punishment. We found his body two hours later. Out of five of us caught in the slide only the two patrollers survived: me and my partner."

"Why don't you mark those spots more dramatically. Something like 'Death Trap.'"

"We thought about that and even had some names picked out. I liked 'Coffin Cove', but it wouldn't matter. Some idiot would always want the glory of conquering something bigger. It's the vacation skier and weekender who are the most common victims of a killer slide. Those of us who live up here ski the back slopes often for a variety of reasons, most often looking for one of you, but we know the terrain. We know the snow and how to read the warning signs. My best advise to you is to stay on the marked trails, and we'll take care of the rest."

Steel Illusions

"I don't think you have to worry about me. After what I saw this morning, I don't think I'll look for that fresh layer of powder with so much excitement."

"Well, I hope the information helped to ease your fears not stir them up."

"Thank's Luke," Laura added. "Actually, I thought my daughter was trying to get a date."

"Mother, I can't believe you said that!" Megan blushed as Laura giggled at her. Poor Luke didn't know if he should laugh or run.

"I'm sorry, Luke." Laura was having trouble controlling her full laughter now.

"Please forgive my mother. She gets a little crazy from time to time. We can only take her out of the special mental home for short trips." Megan shot her mother a knife piercing glare.

"Okay, you've made your point, Dear. But you must admit I got your mind off avalanches." Laura was still smiling at Megan who was beginning to see a small amount of humor in the situation.

"Let's see, fear verses embarrassment. Wow! Tough choice."

Luke was beginning to appreciate the humor of what Laura was trying to do and he couldn't keep the grin from appearing as he added his opinion, "I'd go with embarrassment. It's more fun."

"Not if you're the butt of the joke. Thank you very much, Mother."

"You're very welcome, anytime you need my help, just call. Now may I have my burger before it freezes. Luke, would you like to join us for lunch. I'll try not to embarrass either of you."

The three of them enjoyed harmless and much less embarrassing conversation for the duration of the lunch break before Luke announced he had to get back to work. He promised to meet Megan at the fire pit when he was done with his shift.

R. Z. Crompton

****

The rest of the splendid afternoon Laura spent soaking in the sun and fresh air as she and her daughter went up and down the slopes until notice was given that the lifts would be closing. That was the signal for everyone to head for the large community fire pit behind the hotel. A little Aprés Ski was a favorite way to finish a wonderful day while the majority of skiers crowded the shuttle buses for a ride down the mountain.

"Chelli, I'd almost given up hope of seeing you again. How was your day."

"You'll never believe; Stewart's wonderful. We skied the higher slopes all day. He's a fantastic teacher. Wait 'til you meet him."

Laura couldn't mistake the sparkle in Chelli's eyes. She was definitely in love: again. Chelli kind of fell in and out of love as the sun rose and set. Laura wondered if the girl would ever see past her hormones.

"Am I to assume that you did find your date for lunch?"

"We had lunch, and we're having dinner together. You don't mind do you?"

"Of course not, Dear. Are you bringing him over to the condo or meeting him somewhere?"

"I thought I'd wait until I found out what you're doing tonight. Where's everybody else?"

"They'll be showing up shortly." Before Laura had finished her statement, Krista and Brad came around the corner. "So did you two enjoy your afternoon?" Laura couldn't tell if Krista's cheeks were flushed because of the afternoon sun or if she was blushing.

"Yeah, the upper slopes were magnificent."

"Hey, I skied the top slopes, and I didn't see you. Which runs were you on?" Chelli was quick to question.

Steel Illusions

"Ya know, Krista and I thought we saw you, Chel, but you were so far away. By the way, how was that guy you were supposed to meet?"

Laura saw Krista wink at Brad. She doubted whether the two of them had even left the condo. "How was your lunch, Guys?" Before one of them could answer, Megan came into view with an extremely handsome young man next to her.

Chelli was the first to comment, "Wow! She had a successful mission today. How'd she meet him?"

"Hi. Everybody, this is Luke." He shook hands around the group as he was introduced.

"Mom, Luke got some news about that avalanche we saw this morning."

"Really? What happened?"

Luke had everyone's attention immediately, but Chelli interrupted before he could begin. "What avalanche? I didn't see anything."

"Are we surprised by that?" Brad threw Chelli a quick smile while everyone else nodded in agreement.

"The avalanche was not set off by Mother Nature."

"What? We didn't hear any explosion. Was it set off by the rangers?"

"No. Lots of people falsely believe the rangers practice detailed avalanche control in the back country. There's just too much wilderness out there. People who don't know what they're doing shouldn't be in the back country to begin with. Even the monitoring that's done, isn't done by the rangers."

"It's not?" Krista was the first to blurt out the question that each of them wanted to ask. "We thought they did avalanche control."

"No. Monitoring of avalanche areas, where people maybe easily involved, like over a highway, is done by the Mountain Rescue

-38-

Association, especially since they're the ones who have to do the rescuing. The ski patrol uses small explosives to control dangerous areas on the slopes; but with daily use and grooming it usually isn't necessary. The really dangerous spots are just out of bounds. Otherwise Mother Nature is left to run her own course."

"So what happened today?"

"The best Vail 1 can determine is that a group of kids had been riding snowmobiles back and forth across the face of the slope. The energy released from the machines vibrated through the layers of snow and ice until it eventually shattered, and gravity pulled it down the mountain."

"What's Vail 1?"

"Vail 1 is the top man with Vail Mountain Rescue. He's the person we call when there's any kind of trouble, winter or summer."

"It sounds like the kids were playing Russian Roulette with Mother Nature today," Brad commented.

"A good analogy."

"You know what they say, 'Don't mess with Mother Nature.'"

"You should never mess with a 'Mother'; she always comes out on top."

Everybody moaned at the terrible pun before Laura asked the next question.

"Were there any survivors?"

"So far six machines have been found but only five bodies. I may be called in for the next shift."

"Can't you have a drink with us?"

"Sure. But my hot chocolate has to be served without the peppermint schnapps."

Chelli was still looking through the crowd for Stewart when the waitress approached to take their order. "I wonder what could've happened to him."

Steel Illusions

"Who? Your special ski instructor?" Brad questioned.

"Yes, and his name is Stewart," she answered haughtily. "He was supposed to meet me here."

"Maybe when your lesson was up, he went home to his wife." Megan added.

"Oh, Megan. He's not married."

"I was just kidding. Ya know, he might have gone to help with that rescue. Luke, did any of the other guys go help?"

"Yeah, lots of us go. It just depends on which group he was assigned to. I'll be paged in the next fifteen minutes or so if they need another crew. They won't give up hope of finding someone alive until the last body is recovered."

"That makes me feel better," Megan added to the conversation.

"Are you still worried about being buried in snow?"

"Only when we talk about it."

"Well then we should stop talking about it," Krista told her sister.

"I still want to know even if I'm afraid of it. Maybe I'll learn something that'll save my life someday. So tell me, Luke, how does a person survive an avalanche?"

"You swim."

"You're kidding, right?" Krista's voice was full of doubt.

"Just like you were in the water. As the snow rolls down the face of the mountain, it's like the waves in the ocean; so you swim. The breast stroke works the best."

"Now I know you're kidding." It was Brad's turn to express his disbelief.

"Stewart, here we are." Chelli waved her hand, and everyone turned to watch an extremely tall and tan young man approach them.

"I'm sorry I'm late. I almost didn't make it."

"What happened?"

"They found the last body just before I was supposed to get in the van. Chelli, I'm sorry. I had no way of calling you."

"No problem. At least you got here before we left. If they found the last body, then Luke doesn't have to leave."

"Hey, that's right," Meg added.

"I'm off duty. Waitress?" Everyone ordered another hot drink, some with a little extra boost and some without. A short plump man in lederhosen started playing on his long horn. Low base tones vibrated through the air and added to the atmosphere of the moment.

"Stew, I need some help. The main conversation of the day has been avalanches. Would you please enlighten these city folk about the best way to survive a slide."

"Easy, you pretend like you're swimming and keep your mouth shut."

"Keep your mouth shut? How do you breath?"

"Through your nose. When you open your mouth to breathe or yell, it fills with snow and you'll suffocate sooner."

"Rather than later?" Brad added sarcastically. "A slow death, of course, being better than a quick one."

"I guess it depends on whether or not you want to be found. Believe me if the fall doesn't kill you, you're most likely to die from suffocation especially if your mouth is full of snow."

Now Krista's curiosity was piqued, "I don't understand why people suffocate."

"The victim's warm breath melts the snow right around his face, then the cold air temperature causes the thin layer of water to freeze. Once the ice masks forms, the rescue time is really limited."

"That's right," Luke added. "Keep your mouth shut and try to put your hands up in front of your face. That'll give you more room in front of your face for an air pocket."

"Okay, okay. I've had enough." Megan was finally ready to change the subject.

"Mom, what's for dinner. I'm starting to get hungry. It doesn't take long after skiing all day."

Laura looked at Krista and added, "Well, some of us skied all day."

"Mom! We skied today."

"Sure you did. The rest of us, however, skied ALLLLLL day."

"Laura, that guy is staring at you." Brad motioned his head toward a man sitting on the opposite side of the fire pit.

"Don't be silly. You're just trying to change the subject."

"No. He really seems to be watching you. Have you seen him before?"

Megan laughed as she tried to ease Brad's concern. "Brad, haven't you realized that all men stare at women up here. That's one of the reasons we come. Isn't that right, Luke?"

"Yeah, I guess so, but he really seems more intense than the guys who are just enjoying the scenery."

Laura didn't notice anybody standing around; however, she knew for sure that everyone was hungry. "Maybe Stewart and Luke know a good place to eat. What do ya say guys?"

Luke was the first to speak up, "A lot of us like Cassidy's Hole in the Wall."

"Yeah, that'll work," Stewart added.

"Our favorite place, too."

****

Stewart and Chelli finished up on the dance floor and joined the others upstairs where the conversation had again turned to avalanches.

"Megan, can't you get enough. Poor Luke is going to think all you want from him is his mind."

"It's fascinating. Did you know there are thousands of avalanches every year? But what's really scary is that the most dangerous avalanches can be the little ones caused by skiers or snowmobilers."

"Is that true?" Chelli questioned Stewart as if the others were just trying to scare her.

"Yeah, it's true. People who don't follow the marked trails are more likely to become victims of Mother Nature's wrath. Some of the dangerous spots are marked, but after a big snow or even a warm spell there're so many places where small snow fractures can kill. It's impossible to watch or even mark every place."

"That's enough. Let's order. I'm starving." Brad was tired of the redundant topic.

"Did you see the neat bar stools downstairs? They're shaped like a horse's ass." Chelli was excited about the ambiance of the unique restaurant.

"Yeah, I bet they're about that comfortable too." Brad grumbled as he scanned the menu.

"Don't be so cynical, Brad. I think they're cute. Did you notice the bullet holes in that picture behind the bar?"

"Chel, how do you know they're bullet holes?"

"Mike, the bartender, told me when Stewart introduced us. It was a great idea to come here. Between Luke and Stewart, they know everybody."

"Chel, he was pulling your leg." Stewart laughed.

"No he wasn't. Was he?"

Luke took the opportunity to lighten the mood even further by poking a little fun at Brad. "Krista, your boyfriend may not care for the bar stools, but the men's restroom certainly got his attention."

Steel Illusions

Krista had forgotten the rumors about the men's john. By the way Brad started blushing, she guessed the tales were true. "So, Brad do you want to tell us what happened."

"Not really."

Luke was more than willing to enlighten everyone. "When he got up to the urinal and ah......shall we say, got down to business, he was surprised to look up and find all the people at the bar staring at him."

"What?" Chelli could barely spit out the question through the laughter.

"Well, you'd be surprised too if you looked up and saw some guy watching you take a whiz." Brad looked at the girls as he tried to defend himself.

"Oh, fer sure." Megan giggled.

"He nearly peed on his foot before he realized it was a one-way window he was looking through, and nobody could really see him."

Even Brad couldn't deny the humor as he thought about his shock when he looked up and out the window into the eyes of a beautiful blond. "And Luke here, he just stood back and waited for my reaction. No wonder you volunteered to go with me. I thought maybe you were a little weird or something."

"The only thing weird about me is my sense of humor. I have to enjoy it when I can."

Brad was sitting next to Laura and leaned over to say something. Megan was sure he was poking fun at Luke's pleasure, but he was really more concerned about the woman next to him. "There's a man over there next to the wall who's been staring at you all evening. Do you know him? I think he's the same guy from this afternoon."

Laura tried to look, without being obvious, at the man sitting just off to her right; but the place was packed, so it was difficult to peer nonchalantly through the crowd. As a tall rather plump waitress passed in front of her, she tried once again to get a look at the man. As her line of vision cleared, she found herself looking directly into his eyes. Holding his gaze, she fought the urge to look away. He had a high forehead and long angular face framed with tightly curled coal black hair. The scar on his chin was barely visible under the black stubble. The piercing eyes were as black as his hair, and they threatened her as if he'd been holding her at gun point.

Brad noticed the change in Laura's demeanor. He saw the fear rise in her face as the color faded. Even in the dim light, he could tell she was pasty white. "Laura, Laura? Do you know him?"

"No." The sound was barely audible, and she shook her head as she turned to face Brad when he touched her on the arm. By this time Krista had noticed her trembling mother.

"Mom, what's wrong?"

"That man's staring at me." Laura turned again to look at the stranger, but the chair was empty.

"What man?"

Now Laura felt foolish. Why should a man looking at her be so frightening? She must have been imagining the danger.

"Laura, Laura is that you?"

"Adam, what are you doing here?" A broad shouldered man just taller than Laura swept her off her feet as she stood up to greet him. "It's wonderful to see you. How's Stef and the kids?"

"She and the kids are fine. Hey girls. It's great to see you again."

Both Krista and Megan answered, "You too, Adam."

Laura made the introductions to the rest of the party and invited Adam to join them for dinner. He'd been a long time friend and business associate. Kevin and Laura had even vacationed with Adam and Stephanie. As quickly as her fear had manifested itself in a stranger across the room, it disappeared as a friend stepped in close.

Steel Illusions

Much of the evening was spent reminiscing with the kids laughing at old tales about their parents.

"Laura, will you dance with me?"

"Of course." Laura led the way through the crowd to the dance floor. The noise was so loud there was no way to carry on a conversation, so they enjoyed the rhythm of the music and the company. When the music slowed, it seemed natural to find Adam's arms around her as he moved in close. She was enjoying the cologne he was wearing and the soft touch of his cheek as he whispered something funny in her ear.

The movement in the room slowed in response to the music. Adam's arms tightened around Laura's waist. It felt good to have him close, almost possessive of her after the scare she had felt earlier. Whoever the guy was who scared the daylights out of her had disappeared; and ironically enough, the disappearance coincided with Adams arrival. She not only felt grateful but safe. When the music ended, Laura led Adam back to the table.

"Mom, do you care if we go with Luke to a party in Vail? We won't be gone long."

Before Laura could answer, Brad objected. "You guys go ahead. Krista and I will take Laura back to the condo." Krista jabbed Brad in the ribs, and then she whispered in his ear, "We aren't an old married couple. I'd like to go out."

Laura didn't hear the words, but she understood the communication. "Thanks, Brad, but you don't have to take me home. I can take care of myself. I was doing it long before you were around. You guys go and have fun."

Brad kicked Krista under the table before he insisted, "I'd rather go back to the condo."

Krista didn't understand what was up, but she did know this man well enough to understand there was a good reason for his insistence.

It was Adam who stepped in and solved the problem. "Don't worry. I'd love to take your mom home. I have to go that direction anyway."

Brad wasn't sure he liked the idea, but before he could object Laura agreed to go. Reluctantly, he left with the others.

"Now that they're gone, can we go somewhere for a quiet drink?"

"Sure, why don't you take me back to the condo. We can have something there."

****

Laura poured the smooth amber liquid into the small crystal glasses and walked toward Adam. He was standing in front of the patio door enjoying the view of the mountain as the lights danced on the snow. "I can see why you love this place. It's perfect for a quiet drink. How 'bout a couple of minutes in the hot tube? The steam rising from the water is calling your name."

"You don't have a suit." Laura was disappointed because the idea was a good one. She enjoyed relaxing her tired muscles in the hot water before crawling into bed.

"I don't need one. You go change while I get in. Believe me, in this temperature, I won't linger in the cool out - of - doors."

Good to his promise, Adam was comfortably lounging in the hot water by the time Laura walked through the patio door to join him. Clad in the sleek black and red suit, Laura quickly reacted to the chill of the air by hopping into the soothing warm water.

Steel Illusions

"Wow, this feels so good after skiing all day. I don't know how the kids can stand dancing all night."

"They're young. We did some 'all-nighters' not too many years ago."

"Yeah, I guess we did. No one can say that steel people don't know how to have a good time."

"Have you talked to Kevin today?"

"No, why?"

Adam reached for his drink as he spoke, "Actually, I'm kind of surprised he isn't here."

"Why would he be here? He doesn't ski."

"When I was talking to him the other day, he mentioned that he was looking for some gloves. I just assumed he was headed up here. Why would he need gloves in Tulsa this week?"

"I have no idea what he's up to. Usually he calls first thing in the morning, but I didn't hear from him today. What are you doing here anyway?"

"I had business in Colorado Springs, so I decided to take a couple of days off to enjoy the slopes. You and the girls always talk about how wonderful this place is, so I decided to check it out."

"I'm glad you did; it's wonderful to see you. Are you and Stephanie going to Birmingham for the Globetrotters Meeting?"

"Miss a Globetrotters? Never! Do people still think you're going to a basketball game?"

"Oh, yeah. It's hard to explain that Globetrotters can be steelmakers too. Are you taking the kids?"

Adam laughed at the question. "Our kids at a Globetrotters? Not a chance. Steph would kill me if I even suggested it. That's her chance to relax and have fun."

"I know what you mean. I would've never considered it when the girls were little and now they're too busy. I'm glad they had the chance to go out tonight."

"Me too. I hope they don't come back before I get dressed."

Laura was slow to get his drift. She'd never considered the possibility of being in the hot tub with a buck naked man. "Adam, what do you have on?"

"My birthday suit."

Laura nearly flew out of the water. She had the mental image of a snake slithering across the top of the water. "Jesus, Adam, what possessed you to strip everything off?"

"Don't be such a prod, Laura."

"What if the kids walk in?"

"So what? We've done wild and crazy things before."

"Yes, but Steph and Kevin were with us, and our clothes were on! Even when we went skinny dipping, we left our underwear on."

"Okay! Okay. Hand me a towel and I'll get dressed. I'm sorry. I really didn't think it was such a big deal." Laura turned away from Adam as he stood up reaching for the large white towel. Laura heard him step out of the tub and take a step. She could tell he had taken a step toward her, but she wasn't able to make herself move away. When his hand reached out touching her on the shoulder, Laura nearly jumped out of her skin.

Softly, Adam said her name, "Laura, I........is there any chance that we might?"

Laura, without even looking at him, moved her shoulder away from his scalding touch. It was the only answer he needed. Quickly he reached for his clothes and headed for the warmth of the condo. Laura hadn't noticed the chill in the air; her blood was steaming with anxiety. She couldn't deny her attraction for Adam, but she had always been able to keep it under control. As she stood rigid in the

Steel Illusions

cold, past scenes of Adam raced through her mind. She saw herself dancing with him, kissing him hello. Seemingly innocent innuendoes had been made about sexual desires and future opportunities. Laura was beginning to doubt her ability to separate reality from fantasy. She was still standing in the freezing night air when Adam, fully dressed, walked back to her side.

"Laura, sorry I put you in such an awkward position. C'mon in before you freeze to death."

Startled by the softness in his voice, Laura quickly turned her head to face him. Adam was standing so close she could still detect the fragrance of his cologne even after their brief time in the hot tub. The aroma had been slightly altered by the chlorine in the water. Laura's heart skipped a beat when Adam's eyes locked on hers. He put her towel around her shoulders and lead her into the living room.

"Laura, you know how I feel about you; but you're right. This isn't the place. Someday we'll have time together. Now go get out of that wet suit or you'll be too sick to ski tomorrow."

Laura walked away from him without saying a word. She needed some time to regain her composure. Fingers numb from the cold, she fumbled with the zipper of the red silk sweatsuit. She'd felt like Adam had saved her from her own evil shadows when he'd found her at the restaurant. It never dawned on her that she'd have to keep her guard up while she was with him. She doubted her ability to walk away from him if he touched her again.

Adam was sitting on the couch with his drink in his hand. He'd combed his hair and straightened his shirt. Laura found herself wondering if his proposition had just been in her imagination. She walked into the room and picked up her glass. Nothing more was said about the incident.

"So where are you and the kids skiing tomorrow?"

"We'll want to spend most of the day at Vail. The girls love to ski the back bowls. Later in the day, we'll go up to Beaver Creek. Aprés Ski is best by the big fire pit behind the Hyatt."

"Sounds like fun. Do you mind if I join you. I'll buy everybody dinner."

"Of course, you can join us, and you don't need to buy us dinner." Laura was a bit put off by the ease of his conversation. It was like nothing out of the ordinary had happened. Oh well, what could go wrong if the kids were with them. Everything would be fine if she just made sure the two of them weren't alone. She was beginning to think she'd exaggerated her feelings and Adam's.

"It'd be my pleasure to treat you all to a finely cooked meal. How about Beano's Cabin? I've heard it's an excellent place to enjoy the mountain views at sunset."

"Sounds wonderful. We should be able to see the fireworks from there. I'm sure we'd all enjoy it."

Setting his empty glass on the counter, Adam walked over to pick up his coat. "I'll meet you at the gondola at nine o'clock."

"Great." Laura escorted Adam to the door where he kissed her on the cheek before saying a simple good night.

Laura locked the door behind him. She felt guilty, but she really hadn't done anything wrong. Then she remembered the excitement she'd felt when Adam looked at her and his promise, or was it a threat, that one day they'd be alone. Her first instinct was to call Kevin. The phone rang several times and then the machine gave its mechanical response. It was impossible for Kevin to be in bed and not hear the phone. He must be taking the dog for a walk. Laura left a short message for him to call as soon as possible and hung up the phone. She fell asleep waiting for the phone to ring.

CHAPTER THREE

The strong baritone voice boomed through the telephone receiver, "How'd it go?"

"Fine. Everything's ready. She was easy to find. What about the mill?"

"Almost better than planned. The place is destroyed."

"How 'bout the kid?"

"Well, he's not dead yet. What a fool. He didn't even have his fire proof coat on."

"Why in the hell would anybody go down by the furnace without that coat? Did you see him or talk to him?"

"No. They aren't letting anyone in. If I can't see him, then Bradford can't either."

"Will he tell his wife about the night letter?"

"I'm sure he didn't see it as anything out of the ordinary. Bradford frequently leaves a night letter. The kid's not that bright. Just a stupid Okie."

"And you're an arrogant Yankee asshole."

"And proud of it."

"I still don't see why you coerced the kid into being on the floor anyway. He didn't have to die."

"Die, no. Injured, yes. Besides, I didn't coerce; I merely suggested he make sure all the special scrap got into the furnace. It's his problem if he went down on the floor. However, if he dies, it does help our cause."

"Why?"

"It feeds the plan; and if the kid says anything to his wife, it'll only make Bradford look more guilty. He'll be so busy cleaning up the place and trying to cover his ass, he'll never suspect a thing. So don't start giving me that 'holier than thou' shit. It was your idea to drag

Steel Illusions

Laura off into the wilderness. If she dies, it's on your head. Don't look too deep into my dark soul - you'll see yourself burning in hell."

"Eat shit! You take care of your part of the deal, and I'll carry the diversion one step further. What happens when Bradford puts it all together? He will eventually."

"I'll take care of him and enjoy doing it. What about Laura? She'll recognize you."

"You aren't the only one who can be devious. When you're ready for her to be released, she won't be permanently damaged; unless I'm having too much fun. It just depends on how long you take to get finished."

"Who's doing your dirty work?"

"A local guy who knows the back country and is a specialist in acting without a conscience. How many drivers do you have lined up?"

"One guy on site and three drivers, any more might raise suspicion."

"This is going to be expensive. All those guys will expect big payments. Will we make enough money to pay for this?"

"I really don't care. There's more at stake than breaking even."

"I can't take money out of the company account to pay some guy in Colorado. We have to make enough to pay everybody."

"I don't care how you pay the bill. Take the money out of that secret bank account you set up. It's not like you're destitute. You've made a couple million dollars above and beyond what you've announced to your shareholders."

"How'd you find out about that?"

"Jesus, what else would you do with it, bury it in the backyard? The idea, you fool, is to protect what we've got. If I get canned, it'll only be a matter of time before you're found out too. Stop bitching, you sound like my ex-wife. Stick to the plan and don't fuck-up!" The click echoed through the receiver indicating the end of the conversation.

****

Kevin had spent hours getting people organized for the investigation. After the ambulances and fire trucks had been escorted through the gates, Kevin gave strict instructions that no one else was to get in.

"The press'll be all over the place. You make sure, Harry, that no one, absolutely no one gets in without my approval."

"I understand, Boss."

"Those T.V. cameras won't see anything specific from their perch on the Arkansas bridge."

"Yeah, but eventually they'll end up on the levee, Boss. What do ya want me ta do about 'em?"

"There's not much we can do. They know the view of the damage is better from the levee. It won't be long before the cameras are zooming in on us, but we don't have to let them get up close and personal. The media vultures will be hungry for any tidbit of information they can scavenge. It's their job."

"You're gonna have to face 'em eventually."

"Yeah, I know, Harry. But not now. Have you seen Sam?"

As Harry hurried toward the front gate to stand his watch, he turned to offer the little information he had. "Last I saw, he was making sure all the minor injuries were treated and recorded."

Kevin started for the area where a simple, rustic triage had been set up for the men. He needed to get Sam going in another direction. As safety director, he needed to call OSHA right away.

"Sam. Hey, Sam." Kevin waited for the man's attention before giving the instructions. "Get somebody else to help the paramedics here. You need to call OSHA, and tell them we need somebody out here instantly."

"Joe, call Praxair. We need the oxygen turned off. Pronto." Additional instructions to turn off water to the furnace and the natural gas were given.

Kevin needed to follow a process of elimination in order to determine the cause of the explosion. His first assumption was to check for water in the furnace.

"Dan, I gotta get a look in the furnace. How long?"

"You can getta look in an hour or so, but it'll be some time tomorrow before we can get a good appraisal of the damage. The men gotta clean up those damn mast rollers. Slag was blown into those rollers, and we can't raise the electrode arms."

"We gotta get that roof sitting squarely on the furnace before we can move it. Shit, it's been blown completely off the bezel ring." Kevin rubbed his fingers through his hair as he walked around the back of the furnace. His hand came away filled with singed bits of stubble. It crossed his mind that he wouldn't have to shave for a week. Even if his facial hair did start to grow back, his skin would be too tender.

"Any sign of water?"

"Not yet, Boss. But it's too early ta know fer sure."

"If water was the cause, the refractory'll show signs of hydration." By the time Kevin left the mill, he was pretty sure the explosion had not been caused by a gas or oxygen leak. Water was still a possibility, but the most likely culprit was scrap. Experience told him there were just so many options. Eliminate the obvious and scrap was all that was left. The challenge would be figuring out where the bad scrap came from and how it got in the furnace.

R. Z. Crompton

****

Several hours after the explosion, Kevin was driving along the Broken-Arrow Expressway at a relatively slow speed. The Tulsa skyline was spread out against the glittering night sky. The view Kevin normally enjoyed on his drive home went unnoticed as he mentally surveyed the damage to his meltshop. Somebody was going to be held responsible for this, and Kevin knew that until he had definite proof, the company would not only hold him responsible, he'd most likely be the scapegoat. Waterman, as V.P. of Production, would certainly cover his ass by blaming Kevin as soon as possible. It was ironic that Waterman was the one who hired Kevin and convinced him that Tulsa was such a wonderful place to bring up children. Well Tulsa was a wonderful place, but under Waterman's dictatorship, the mill was a snake pit. Kevin had been at odds with the man since the beginning. Waterman couldn't afford to just out and out fire him because Kevin had set too many records, improved the production, lowered the operating costs. Every man's bonus, including Waterman's, had gone up since Kevin had taken over. Now Waterman might have the reason he'd been waiting for.

As Kevin turned off the B.A. Expressway, he mentally logged his schedule for the next few days. He needed to try and get a statement from Pete. Pete would know more than any one else about the heat that was being made. Kevin hadn't ordered any special grades, so there was no reason for any unusual scrap. There had to be a reason Pete was down on the floor.

It wasn't until the wind nearly jerked the steering wheel out of Kevin's hand that he focused on the highway again. Thank God for the good old Oklahoma wind sweeping down the plain, Kevin thought as the words of the song spilled from his lips. The gust of pure energy had just kept him from driving off the road. After the visit to the hospital, Kevin decided he'd go home and shower. He'd pack a bag and take the dog to the vet for a couple of days and then head back to the mill. Good old Chance would be happier boarding with people who could cater to his needs rather than home alone. Kevin didn't know how long he'd be at the mill, but it'd be until he had some answers.

Kevin locked the door of his red mustang and headed for the hospital entrance. Visiting hours would be over soon, and he had to talk with Pete if it was possible. Kevin knew which floor to head for without having to ask. Visitation would be limited to family only, so he'd have to convince Amy to let him see Pete. Facing Amy would be the hardest part of this whole mess.

Bright lights flooded the elevator when the doors opened. The ethereal whiteness of the hallway not only symbolized cleanliness, it gave an uncanny illusion of being heavenly. It was like so many movies Kevin had seen where somebody died and then floated toward the bright lights of heaven. Kevin knew the sterility served a purpose, and he wondered if the heavenliness didn't also serve a purpose: most of the people in this ward left spiritually before they did physically. Kevin proceeded down the hallway with a heavy heart.

"Excuse me," he said to the solemn looking nurse sitting at the nurse's station. Her light brown hair was severely coiffed, but her eyes and smile portrayed a softness of character. Kevin wondered which trait belied her true nature.

"Excuse me, Miss. I need to talk to you about Peter Hayes."

"Yes?"

"How's he doing?"

"He's in and out of consciousness. His wife has been with him all evening."

"I really need to talk to Mr. Hayes about what happened."

R. Z. Crompton

Nurse Reynolds immediately went from simple receptionist to protector. "Not a chance, Sir." The "Sir" held more of a mocking tone than any respect. "Only his wife is allowed to see him."

"Would you please tell Mrs. Hayes that Kevin Bradford is here?"

"You can't see him." She responded impatiently making it clear that even a wife couldn't override her authority.

"I understand, but I have to find out as much as I can about the events which lead to the explosion. If Mrs. Hayes can ask him a couple of questions, it would help me find the cause of his injuries.

"Okay, I"ll tell her, but she may not want to talk to you." The nurse marched away before Kevin could respond.

Kevin was not emotionally prepared for Amy when she walked out of the room. Tear ravaged eyes were filled with hatred and anger.

"You have a lot of nerve showing up here and asking to see Pete. What do you want to know? Has he said anything? Anything that would point the finger? Has he blamed you?" She spat the words at his feet.

"Amy, I'm sorry. I know how you feel."

"You don't know anything. Pete looked up to you. He'd do anything you asked. Now, he's gonna die for you."

"I don't know what you're talking about. I didn't tell him to do anything."

"He's only said two things to me, 'Bradford did...' and 'tell Bradford...' that's it. Four little words. Not 'Honey, I love you' or 'Honey, I'm sorry,' it was only your name he said. Go to hell, Mr. Bradford." She turned and walked away.

Steel Illusions

****

The high pitched sound from the small black box echoed through the empty house, but no one was there to hear the page or answer the phone. Kevin was on his way back to the mill. He only had a few hours to come up with some kind of evidence for Waterman. Kevin's conversation with his boss earlier in the evening had not set well; if he couldn't find reasonable explanations as to the cause of the explosion, his career was over.

Anthony Mason would be his best ally. As President of Green Country Recycling, Mason needed the steel mill up and running. GCR was the primary supplier of scrap to the mill; and if the mill was down, so were they. Kevin had reached Mason through his answering service, and he had agreed to meet with Kevin as soon as he could return from Chicago. Together they'd go over the incoming scrap shipments, looking for anything that might be suspect.

With his mind racing through possible calculations, Kevin pulled off the exit toward the mill. He needed to make sure the day crew had gone home. They'd be exhausted, and that's when mistakes were made. After a few hours rest and a good meal, they'd be more valuable to him. As he pulled into the parking lot, a deluge of camera men and reporters crowded around the car.

"Mr. Bradford, can you give us a statement?"

"In the morning; the company will issue a statement in the morning." He yelled through the glass of his car window. That response wasn't good enough. Reporters bombarded him with questions as if they were throwing stones. One question did catch Kevin's attention.

"Was anybody killed?"

He had to think for a second before answering. If he said *no comment*, they'd certainly assume the answer was *yes*. If he said *no,* he wasn't lying. In fact, that might take some of the pressure off. "No. Nobody was killed," was the only information he gave before pulling through the security gate.

In the morning after he'd had a chance to talk with Waterman and the President, an official statement would be given to the press. It wasn't Kevin's responsibility or even his right to be the spokesman for the company.

Kevin ran his fingers through his hair again as he rounded the corner to his office. The man sitting at his desk was dressed in dusty blue jeans, a faded flannel shirt and mill coat. The yellow hard hat and safety glasses finished the outfit. However, it wasn't the clothing that made the worker, and Kevin recognized the imposter immediately.

"Rick, how the hell did you get in here? And what are you doing at my computer? You belong outside with the others."

"I was enjoying your screen saver. Adam's Family? Cute. The cawing sound of the crows caught my attention."

"How'd you get in?"

"I can't give away my secrets any more than you can."

"There's nothing I can tell you. The company'll make a statement in the morning."

"Sit down, Kevin. I know you and your ethics. I also know something you don't."

"Yeah, what?" Kevin sat down behind his desk when the reporter vacated the chair.

"Didn't you get my message?"

"When'd you leave it?"

"About an hour ago, I called the house. Told you I needed to talk to you right away. Also called your beeper."

"Shit." Kevin patted all his pockets. Then he remembered leaving his beeper on the bathroom counter while he took his shower. He always left it by his car keys; but in case someone needed him

Steel Illusions

immediately, he'd taken it into the bathroom where he could hear the damn thing go off. "Shit! Shit! Shit. I hate it when I forget that thing. Oh well, I wouldn't have returned your call anyway."

"Sorry to hear that, my friend. Why do you have an answering machine and beeper if you don't return the calls?"

"I listened to half a dozen or so messages; they were all from the press wanting interviews, so I got into the shower. Rick, I can't tell you anything; friend or no friend. I can't."

"I'll make you a deal. You tell me how badly the kid's hurt, and I'll tell you what his wife is saying to anyone who'll listen."

"Jesus Christ! I never even considered her talking. You guys tracked her down at the hospital?"

"Don't try to pin this on us. She called the station and requested the interview; she wanted a TV camera and crew to see her at the hospital. I went without the camera just to see what she had to say. Hysterical wives don't always give the most credible stories. She must'a called right after you left."

"You! Why'd you take it."

"I was the lucky one who answered the phone. Lucky for you that is. I tried to call you before I left, but after what she said, I couldn't leave it to chance that you'd eventually get back to me."

"Okay, I owe you. You wouldn't have gone to so much trouble if it wasn't serious. So what's she saying."

"She's saying you're responsible. If her husband dies, you're the one who killed him."

"My God. I knew she was angry, but I had nothing to do with the accident. At least, not intentionally. I swear, Rick. I'm trying to find out what happened, but there was nothing ordered that could have caused such a violent reaction."

"I can sit on this for a while, but she'll be calling the paper and anybody else who'll listen. The longer you wait to give a statement, the more people will think you're trying to hide the truth."

Kevin paced across the small room. Rick was right. He needed some answers quickly. "Thanks for letting me know. I'd hate to see my name spelled out for murder on the headlines of the morning paper."

"She thinks she can sue the daylights out of you. You need to have your company attorney ready for her."

"She can't sue the company. We didn't do anything wrong."

"If she can prove or even make a good case for blatant neglect, you'll be in trouble. Even if she can't win a case, she can tie you up in court and make your life miserable. Kevin, she can ruin your career."

Kevin laughed as he practiced the nervous gesture of running his fingers through his hair. "Rick, if I can't find out exactly what happened, my career's over no matter what she does."

"Is there anything I can do to help?"

"As a friend or as a reporter? Sorry, I didn't mean that the way it sounded. You wouldn't be here if you weren't a friend. You would'a just run the damn story." Kevin stared at the picture of the Adam's Family on his computer screen. "Ya know what's funny is that Pete's probably the only one who can give me some of the answers I need, and she won't let me talk to him."

"Maybe she'd let me." The soft spoken man speculated. His honest, handsome face gave Kevin some hope.

"If she talked to you once, maybe she'll give you some of the information I need as long as you don't tell her what you're up to."

"No kidding. What do you want to know?"

"Ask him or have Amy ask him why he wasn't wearing his flame retardant coat, why he was on the furnace floor, where'd the scrap

Steel Illusions

come from? Any information from him might help me. Rick, there's a reason why that kid's hurt so badly, and only he knows what it is."

"I'll do what I can. Now, are you going to answer your part of the exchange?"

"What question? About Pete?"

"Yeah, how bad's he hurt. His wife won't be very cooperative if her husband dies. How much time do I have?"

"Are you busy right now?"

"That bad? It's just after midnight. She won't talk to me now."

"It's bad. Pete's seen burn victims out here before; a doctor doesn't need to tell him how bad he's hurt. He knows, if he's been conscious at all."

"Okay, I'll tell her I needed more information in order to get the story on the morning news. That doesn't give you much time, but it'll help. When I talk to her again, I'll call you. Where can I reach you, or a better question is where can I leave you a message you'll answer?"

"Smart ass. I'll be here, but don't call my beeper." Both men smiled to relieve some of the stress. "Just call my office; if I'm not here, the secretary will answer."

"What's the best way out of here?"

"The same way you got in: with your secrets, or I could take you out the front gate, but then all the other reporters are going to know you were in here."

"I'll see myself out."

"Good, I don't want to know how you do it. Rick?"

"Yeah?"

"Thanks. Next time I'll answer my messages."

"Good. And you're welcome."

Kevin sat down at his computer and pulled up the various scrap deliveries over the last two days. There was nothing he hadn't ordered, nothing unusual. Now the real investigation started. He'd have to try and re-create every shipment from point of origin.

"Hey, Boss?"

"Yeah, Dan, I want you to make sure all the day crew has gone home. By midnight the next shift should've all checked in, so tell the three-to-eleven shift to hang on a few more hours. When the day crew comes back in the morning, keep half the midnight guys on and send the others home. We'll have to do some serious overtime until we get this place cleaned up."

"Already taken care of. Only day crew still here are the general foremen."

"Good. How's it going out there?"

"We should be able to get a good look at the inside of the furnace in the morning, probably by eight or so. Personally, I don't think it was water. I found several pieces of scrap just like this one." In his hand, Dan held a curved piece of scrap. "Any fool can tell by the shape of this that a cylinder of some kind was in the furnace. I'd guess by the number of pieces we found, there were probably several in the last load of scrap; and by my guess it was a load of #2 Heavy Melt."

"That's not proof. We use lots of scrap that used to be part of closed containers."

"It may not be proof, but it certainly shows the potential."

"Yeah, that it does. Make sure the guys save all the pieces. The Fire Marshall might be able to tell us more tomorrow. Now you go home and get some rest. I'll see you in the morning."

Dan was a good, loyal employee. He'd use every possible angle he could in order to get the truth. Like a lot of the others, he needed the mill to provide him with a living. Since the mill was the main source of employment for the small community about seven miles west

Steel Illusions

of Tulsa, Dan had worked nearly his entire life making steel. And now, it wouldn't be long before he could retire.

Kevin was thankful he'd had Dan as his general foreman for the last few years. The man would be sorely missed when he did decide to take retirement. God didn't make his kind of true blue any more. He'd weathered the worst recessions, the bad managers, and the terrible accidents just so he could enjoy the good times in a town he loved. In a society where employees change jobs with the seasons in order to follow the money tree, a few loyal men stay with one company until retirement. Their reward isn't just the monthly check, there's a sense of pride and accomplishment. Kevin had a special respect for Dan and a few of the others who had spent their adult life with one company. These men were the true owners of the company not the guys in the corporate towers on the east coast.

"Hey, Kevin, you in here?"

"Anthony, I'm glad you're back. How'd you get here so fast?"

"Kevin, its nearly dawn. How long have you been sitting here?"

"I've been going over the records, but there's nothing here. I need you to check with your scrap sources, see what's been coming in. Dan thinks bad scrap caused the explosion. I should be able to get into the furnace shortly, but I don't think I'll find any hydration in the brick. By the way, thanks for coming back so quickly."

"No problem. If you're down, I'm losing money; and I don't like that. How's the kid?"

"Last I heard, he was still hangin' in there. His wife is blaming me and yelling 'murder' to the press."

"She wants a person not a company to be responsible. It's easier to hate a person. You were Pete's boss, so you have the honor of receiving the blame. What about the crews?"

"I sent most of the day crew home before midnight; when they get back, I'm going to run through everybody's steps just before the explosion. I have to meet with Waterman in an hour or so; I was hoping I'd have something concrete to tell him. "

"Here's some breakfast. A mug of coffee for me, a Mountain Dew for you and some bagels. Have you eaten since yesterday?"

Kevin shook his head. "No appetite."

"You won't be any good to yourself or the mill if you get sick."

"Anthony, a Mountain Dew isn't exactly beaming with nutrition."

"No, but the bagel is, so eat and let's get to work. I brought my delivery sheets with me. Let's make sure my records show what's leaving our scrap yard is the same as what's coming in to the meltshop. Have you checked with your other suppliers yet?"

"American Recycling has been shipping us loads of beach iron, and our heats yesterday didn't call for any beach iron. I'm still trying to get in touch with Adam from United Metal Services, but he's somewhere in Colorado."

"Isn't that where Laura and the girls are?"

"Yeah. Ya know, I haven't even called her yet. I imagine she's left a bunch of messages for me too. I'll get her later."

Kevin continued after he got his mind back on track, "United Metal Services's shipment came in as shredded material the day before yesterday, and we used it in the morning. My records show #2 Heavy Melt coming in from GCR."

"That's what I show too. What about your scrap inventory on the ground? If the scrap had been dumped into one of your piles and then reloaded, you'll never be sure where it came from."

"That's what I'm worried about. But first I wanted to check all the incoming railroad cars for the last forty-eight hours just in case there's something obvious. Every load Sam checked-in matches what I ordered and what's been logged in the computer."

"What about all those cars belonging to Sand Springs Railroad? Could there have been anything left in one of them?"

Steel Illusions

"Not likely, and certainly nothing we could prove. I'm hoping the Fire Marshall might be able to provide me with some expertise."

"In what way?"

"When my house burnt a few years back, he was the one person who could tell me what burnt first and why. I was hoping he'd be able to tell if a particular piece of scrap was responsible for the explosion."

"That's a long shot, Kevin."

"I know, but those guys can do it in the movies; there's got to be an expert out there who can give me some information."

"Yeah, right."

Sam walked into the office after a quick knock on the side of the open door. "OSHA is here, Boss."

"Thanks, Sam. Show 'em around."

"Dan's back and said you could get a pretty good look at the furnace if ya want."

"Be right there. Anthony, I gotta get going. If you come up with anything, let me know." Kevin vacated his chair and headed for the door.

Sam lead the way out of the office and across the yard. The sun was just coming over the horizon painting the sky with thin ripples of pink and red. A light north wind whispered through the trees promising to keep a chill in the air all day. Kevin didn't notice; he only saw reporters standing at the gate and cameramen focusing zoom lenses on anything they thought was interesting.

"Dan, what are you doing back here so soon? You're supposed to be getting some rest."

"My furnace. I wanted to be here for the first look."

"So, what do ya have?"

"It's pretty bad, Boss. The electrodes were shattered. They're laying in pieces on the bottom of the furnace."

R. Z. Crompton

Kevin could see the damage. He looked into the ten foot deep furnace filled with the shattered pieces of graphite. Fortunately, most of the water cooled panels which were dislocated were not leaking. Punctured panels would have meant water damage. Water and steel: it was like trying to mix oil and vinegar: one way or another, they'd separate. Water was a necessity for making steel, but the very water you had to have could be the cause of a violent explosion if it accidentally mixed with the molten steel.

The winch cables, which control the movement of the electrodes, were broken and two of the three mast columns were badly bent. One of the electrode arms, which carried the electrical current, had been completely sheared off and was laying on the top of the furnace. This furnace would be out for days.

"There's no signs of hydration, Boss. No places where the brick turned muddy or powdery. Even if water had been leaking behind the brick, there'd be signs; and there's nothing, not even the water cooled panels are leaking."

Kevin stared at the black walls of his #1 furnace. "We need the charging crane to get those electrodes out of the furnace."

"That'll take a couple of days." The crane, which rode on a rail system approximately sixty feet in the air, was blown off its track. "We'll have to get some jacks up there to get it straightened out, but I don't know how bad the electrical damage is."

"Let's get electrical and mechanical to check out the #2 furnace and 9 West charging crane. If #2 furnace and 9 West are in working order, we can start up as soon as OSHA gives us the 'okay'."

"We gotta get 9 East crane out of the way. It'll be two or three days before we can get it working again."

"Try to get it jacked up on its rails, so we can use 9 West to push it down to the east end of the building. That should clear up the rails so 9 West can move the slag pots in and out while #1 furnace is down."

"I sent Bob and a couple of others up to check out the bag house."

"Good. I know OSHA won't give us a clean start up until we can prove nothing toxic is floating out of this mill into the atmosphere. I'm not as worried about the bag house as I am about the drop-out box for the fourth hole evacuation system. If there's any damage, it'll be in that part of the process."

"Yeah, we checked it already. Some minor damage, but nothing we can't fix up without much problem."

"Our priority for parts and maintenance is 9 East crane. We gotta have that crane working before we can clean out the furnace. I'll be up in Waterman's office for a couple of hours."

CHAPTER FOUR

At nine o'clock Adam was waiting at the steps of the gondola for the five skiers to make their appearance. He was surprised to find himself staring at two women in a group of five. Both were tall with long legs magnified by tight fitting black ski pants. Long auburn curls glistened in the sun and the ruby red lips moved with rapid conversation. Almost instantly, Adam realized he was watching Laura and Krista; but from a distance he couldn't tell them apart. It wasn't until Laura took off her sun glasses that he was sure which one was the mother and which one was the daughter. Funny, he'd never noticed the uncanny resemblance. Krista was a little girl when he'd met her; and over the years, he'd just never noticed the similarity. Krista was the mirror image of her mother as they approached. Even their stride was the same. For the first time his physical craving for Laura made him feel slightly guilty, but only slightly.

"Good morning, Adam." Laura was curious about the look on Adam's face as she spoke to him. "Is something wrong?" His handsome face portrayed a bewildered look. Laura admired the dark eyes and soft wavy hair. There were no goggle lines telling tales of past ski adventures. Laura knew Adam's past was full of colorful stories. It didn't matter which city, metropolitan or minuscule; there was always excitement waiting for Adam and his entourage.

"No, I just hadn't realized how much you two look alike."

"I'll take that as a compliment, but I'm sure Krista doesn't."

"Sure I do, Mom. It means I have a few good years left." Everybody laughed and headed up the stairs to the gondola.

As they waited in line, Adam questioned Brad about the evening before. "So, did you guys have fun last night?"

Steel Illusions

Megan jumped to answer before Brad had a chance. "It was wonderful. We danced and listened to ski stories until way after midnight. Mom was sound asleep when we got home."

Laura didn't miss the wink Adam gave her. She felt her heart skip a beat. Instantly, she realized that she was supposed to have been Adam's adventure for the evening. She didn't know whether to be insulted or flattered.

The line moved fairly quickly, and soon they were loading their skis on the outside of the gondola. Adam took his stance close to Laura as the skiers filled in the empty spaces. When the gondola began its journey up the mountainside, Laura noticed the large puffy clouds floating over the peaks. Yesterday the sky had been completely void of the white puffs.

Krista was the first to offer some real information about the evening. "Mom, you should have seen Chel. She didn't stop moving all night."

"I was just enjoying the company," Chelli defended herself.

"In term's of male flesh, you were working up a hell of an appetite," Brad added his observation.

Everybody laughed at Chel's defensive look. Her hands, if she could have moved them, certainly would have been on her hips in a defiant gesture. "I was simply enjoying the buffet that had been spread out before me."

"What were Stewart's feelings about your wandering appetite?"

"Actually, he said he was glad I didn't smother him. If you noticed, he didn't exactly sit around pining away for me."

"True. True."

"And he did give me a ride home."

"A loooong ride home. It certainly took you a lot longer to drive that stretch of interstate than it did us."

"Yeah, but I wasn't the last one home." Everybody looked at Meg, who blushed deeply at the accusation.

In an attempt to recover some dignity, Megan answered, "Was I? I didn't realize."

"Yeah, right." Brad was the first to retort.

"Do you always have to be so cynical? It really gets old." Megan looked directly at Brad.

"Your dad made me promise to keep you all in touch with reality. I'm just doing my job."

"Krista! He belongs to you. Take care of him. If we wanted Dad's cynicism along, we could've brought HIM on vacation."

Krista jabbed Brad in the ribs in an attempt to get him to shut up. She could tell by the look on her sister's face that this was not a joking matter. Brad took the good natured warning and put his warm, handsome smile on his face before he offered his next sarcastic remark. "So you really like this guy, huh?"

At this point, everybody unanimously chastised, "Brad!"

"I can't help it. So, Chel where's your Prince Charming today? I thought you'd be taking another private lesson." Brad couldn't erase the smile on his face.

"So now you're gonna pick on me? He was already booked for a lesson, so I figured I'd make sure you and Krista didn't get lonely today."

"OOOH. He's got another woman already, Chel. Aren't you jealous?"

Chel smiled as she wound up for her response, "Oh yeah. I'm jealous of three eight year old Japanese girls. He'll be at the fire pit waiting for me by four o'clock."

"Are they always like this?" Adam questioned.

"Never a dull moment," Laura responded as the gondola began to swing with the motion of the wind.

Steel Illusions

"I hate it when the wind blows these things all over." Megan grabbed on to Laura in an attempt to steady herself.

It was Brad who once again fed the conversation. "Did you ask Luke how they rescue people from swaying gondolas?" As Megan's complexion became more pale, they all knew it was best to keep her talking.

The rocking motion continued, and Megan felt her stomach flip flop. Her hands were becoming clammy. She turned around so her line of vision followed the same path as the gondola and continued the conversation. It was the only distraction that would keep her from puking on everyone's feet.

"No, but I should have. I wonder when one of these last fell to the ground."

"Megan, stop being so morbid. You should've heard her last night, Laura. You'd have thought she was planning to rescue somebody today." Brad stated.

"Maybe I will. At least I feel better knowing I could if I had to."

Brad gave a gallant nod to Megan. "I feel better knowing my life is in your hands."

"Yeah, right. It just seemed like the natural conversation after all those guys got back from the snowmobile accident. Mom, this one man told the best stories."

Now the others were excited about the conversation. "Yeah, Mom. You'd a liked him. His name was Tim...Tim something. He was in charge of the search."

"It was heartbreaking when he told us about the young man who died just because he wanted to try snow boarding," Chelli added.

"What happened? Snow boarding isn't that dangerous." Laura wanted to make sure the conversation continued. It was still a long way to the top. Everybody, including Brad, let Megan do the talking.

-74-

"It's so sad. Newlyweds came to the mountains for their honeymoon, and the guy wanted to take one radical snowboard ride."

Then Chelli jumped into the story. "Yeah, Tim said the man wanted a ride just like he'd seen in the movies." Then she took one look at Megan's pale face and let her continue talking.

"Unfortunately, he didn't know anything about avalanches; so when his wife dropped him off along the side of the road, she just drove away thinking she'd pick him up in a couple of hours."

In the excitement, Krista had to add her part of the story, "When she went back to get him, he wasn't there. Tim said he knew right away what had happened. The guy got caught up in an avalanche. The man was in a high risk area and didn't even know it. Isn't that a depressing story?"

"I think it's kind a scary," Megan threw in.

"Actually," Brad stated, "I preferred the story Sande told us about the naked skier." That statement grabbed the attention of everyone in the gondola.

Adam was the one to ask for more details, and Brad continued. "Some lady needed to take a leak, but she wasn't near any restrooms. When she came to a big group of trees, she just eased her way in and dropped her drawers. Unfortunately, she didn't take off her skis, and when she squatted she lost control and started skiing down the mountain."

"You're kidding." Laura laughed.

"She went screaming down the mountain with her arms swinging her ski poles uncontrollably in the air."

Megan laughed and continued, "Imagine looking over and seeing a 'full moon' on skis pass by."

"What happened to her?" Adam questioned.

"Eventually, she ran into a tree. The ski patrol had to come up and get her. She was banged up pretty good because they couldn't

Steel Illusions

move her. She had to lay there almost naked until help arrived." Megan explained.

"How embarrassing!"

"Sande said some poor guy had been brought into the emergency room with a broken arm. He created quite a picture when he relayed the cause of his injury. 'It was the craziest thing I'd ever seen. A naked screamin' lady came skiing past me. Never seen anything like it. I ran right into a fuckin' tree.'"

Everybody in the gondola broke out into full laughter.

"Oh, but that's not the really good part." Brad could hardly speak through his laughter. "The naked lady was lying in the next bed and heard everything he said."

"Oh my God. That poor woman, she must've been so embarrassed." The laughter carried them until the gondola came to a rest at the top of the mountain. Laughter had taken care of Megan's queasiness; the color had returned to her face and her eyes danced with laughter.

Adam followed so close to Laura she could smell his cologne, and there was no doubt it was his. She'd enjoyed it for years.

"Did you ever hear from Kevin?"

"No. I wonder what's going on. It's not like him to go without calling. Actually, I'm getting kind of worried."

"Maybe he's coming out to surprise you."

"It'd be a surprise all right." As the group moved out away from the crowd, Laura continued her conversation. "By the way, did anybody check the weather report? The flakes are flying."

"I assume you mean snow flakes as opposed to the female skiers in general."

"Brad, you're hopeless."

"No, he's terrible." Megan was beginning to get tired of the constant smart ass remarks.

Adam tried to put the comment into context, "Actually, he's probably right."

"Adam! You're as bad as he is!"

"We appreciate the flakes; both types." Adam grinned.

"The weather, please." Laura requested.

Brad tried to redeem himself by giving his version of the local weather forecast, "Oh yeah, wind and snow."

"No shit, Sherlock. Any flake could've figured out that much."

"Really, that's all I heard."

"I'll ask Luke at lunch. He'll know." Megan's tone of belittlement was not missed by Brad, and he knew he deserved it, but he couldn't miss the chance for a good shot. He didn't trust any man who lived simply according to the snow fall. Kevin wouldn't have wanted some ski bum to spend too much time with his daughter.

"Ooooh Luke. He knows everything."

"Shut up, Brad. You can be such a pain in the..."

"Asset.....I'm an asset to this group because I keep each of you from getting too full of yourself."

"Krista, he doesn't get to come next year."

"Brad, if you want to leave this mountain alive, be nice." Krista swatted at his butt with her ski pole and quickly skied away.

Brad wanted to make sure Laura wasn't mad at him, so he threw out the question, "When and where is lunch?" A positive answer would mean he was still part of the family. Even though he had no real doubt, it was best to make sure.

"Two Elk Lodge. Eleven thirty."

"See ya there." Brad skied after Krista toward the rope tow which would carry them to the black diamond slopes of the Siberia Bowl area.

At the top of the mountain, it was easy to see the front moving in. Off to the west, the white fluffy clouds were no longer floating harmlessly across the sky but became an ominous gray line of frozen precipitation which was just beginning to creep in around the peaks of the opposite range. If they were going to enjoy the time they had on the slopes, they'd better get started.

"C'mon, Mom." Megan yelled as she and Chel headed out across the snow.

"I'm right behind you."

"Laura, how difficult are the runs back here?"

"Adam, you should've asked that before we got up here."

"On the way up, it didn't seem like a big deal; and I didn't want to sound like a weeny. Now that we're up here, it seems overwhelming."

"Don't worry, you'll get down in one piece." Laura loved the wide open area of the China Bowl. There wasn't nearly as much traffic as on the front side of the mountain, and the terrain challenged her without the moguls which tended to destroy her knees.

Adam followed her to the top of the run and was overwhelmed as he looked over the vast beauty of the arena. "So this is a back bowl. Awesome!"

"Yeah, it is," was all Laura said before going over the edge of the slope.

****

Snow flurries intermittently blurred the vision of the trek down the ski run; unfortunately, the brief interludes of sunshine were becoming shorter as the morning wore on. The snow increased, and so did the howling, bitter wind. By lunch time, fingers and toes were numb. Laura could hardly wait to get indoors. She and Adam assumed they were the first ones to give in to the elements; but the lodge was already filling up. They got something hot to drink and looked for an empty table where they could wait for the kids. Off in the corner, Laura picked out the profiles of two familiar heads. Laura could tell they hadn't been inside very long because their hair was still dripping from the melting snowflakes. Both girls were trying to wipe the water off before it damaged the carefully applied eye make-up. Their cheeks were chapped red from the wind and cold. Chelli and Megan were so deep in conversation they didn't see Laura and Adam approaching. Megan's confession was out before she realized her mother was within hearing distance.

"I think I'm in love with him."

"Excuse me?" Laura's remark was requesting an explanation not a chair.

"Oh, hi, Mom. Yeah, you're excused. What'd you do?"

"I didn't do anything, but I'd like to know more about this guy you're claiming to be in love with."

"Me too." Adam added pulling up a chair and looking tentatively at Megan.

"I don't know for sure; I just met him."

"My feelings exactly."

"Oh, Mom. Don't exaggerate. Besides, you know Daddy claims to have fallen in love with you at first sight. If it can happen to him, it can happen to anybody."

"Yeah, Laura." Chel came quickly to Megan's defense. "You always talk about how you two only dated for six weeks before you got engaged. And look how happy you've been all these years."

Adam turned his attention and mocking smile to Laura. "It's your turn."

"With you around, we don't need Brad."

"Did somebody say my name?"

Steel Illusions

"Oh great! This is all I need." Megan was feeling cornered as Brad and Krista joined the group, putting a tray of stuffed baked potatoes on the table.

"What's up, Meg? You can tell us. We always find out anyway."

Chel, as usual, was the one to broadcast the news. "Megan's in love with Luke."

"Chelli!" Megan admonished the young woman sitting next to her.

"What? He's right. They always find out anyway." Chelli tried to defend herself.

"Meg, let me give you a piece of advice."

"This outta be good. Are you giving it as my surrogate father or is it just from you?"

"I'm sure it's from both of us," Brad said earnestly.

"In that case, I can guess what it is: this guy is only after one thing. He's just a ski bum."

"Well that's about it. If you already know the speech, how can you possibly want to be in love with him?"

"You don't always get what you want." Megan's anger flowed in her words. Adam and Laura looked from one speaker to the next wondering which one would win this verbal battle. "Just look at poor Krista. She's stuck with you whether she likes it or not."

"Wow! Touché, Megan."

"I'm not stuck." Krista wasn't sure if she was defending Brad or herself.

Laura decided she'd better interrupt before Megan stormed off without eating any lunch. "Megan, you're an intelligent young lady, and you've worked hard to get where you are. I think you know which questions to ask yourself. You don't know much about the man, but your instincts will serve you well. They always have."

"That's part of the problem, Mom, and everybody else." Megan smiled at the people around the table. They had all been friends for years, and Brad was right: there were not many secrets among them.

"What do you mean?"

"My head and my heart are telling me two separate things."

"For now, listen to your head because it's not your heart doing the talking; its your hormones." Brad offered the advice as a brother rather than a sarcastic friend.

Megan blushed beyond the red chapped cheeks and turned to look out the window.

Now it was Krista's turn to offer some advice. "He's right, Meg. You can satisfy your hormones without giving up your life."

"How in the hell would you know?" Brad was instantly looking at Krista waiting for an admission of philandering from the woman he loved.

The gleam in Krista's eyes and the smirk on her face absolutely stated that she wasn't giving up any secrets. And that left everybody at the table wondering how they could've missed something so exciting over the years. All eyes stared at Krista as she calmly raised her napkin to her lips, blotted them clean, and excused herself to the lady's room.

"I think I'll go with her." Chelli was instantly up and running off behind the suddenly mysterious Krista.

"Megan, we'll only be here one more day. Why are you so upset about this?"

"Luke asked me to stay over."

"What?" Laura's first response vibrated with negative tones. Then she collected herself and addresses her daughter again. "What a....was your answer?"

"Nothing. I want to say 'yes'. This whole thing is so frustrating. When you were my age, the fear was getting pregnant. I hate it. The girl is still the one who has to say 'no.' The responsibility is still ultimately with us. Now getting pregnant is nothing compared to the terrible diseases that flow from one person to another, destroying lives." Megan rambled on about her dilemma while the three at the table simply stared at her.

"I'm afraid you were right, Brad. I hate to admit it, but I agree; you guys can really be scum bags."

"Gee thanks. I think."

"Megan, are you talking about staying for the weekend or for good."

"I'm not completely crazy. Give up school for a weekend fling! Not a chance. I've worked too hard and too long. But I'd like to forget about all the rules and fear for a couple of days and enjoy the desires of my body."

Brad's curiosity was piqued and the next question slipped out before he remembered who else was sitting at the table. "Meg, are you still a virgin?"

"Why do you think Chelli jumps all over the place. She's as scared as I am; she just never lands long enough to make an impact."

Brad wasn't quite speechless, but he was surprised. "Shit! Chelli's a virgin too? Wow! I missed that one."

Laura and Adam didn't know what to say. The conversation was moving faster than either of them could respond. Megan barely acknowledged Brad's question before continuing her harangue about the double standard of society.

"You know what I mean, don't you? I'm sick of guys putting us in this position. We get blamed if we do and blamed if we don't. Do you know how lucky you are, Brad? You don't have to worry about whether or not Krista had a wild weekend or not..."

"I do now!" Brad blurted in, but Megan stopped only long enough to take a breath.

"I think about where the guy has been every time I go on a date. I hate it! Some really fine guy will be kissing me, caressing my arm and.....well, at some point I ask myself 'is this guy worth dying for?' Boy, that's a mood breaker for sure. I say 'good night' and go in to take a cold shower."

"So ask yourself the million dollar question, Megan," Brad suggested.

"I did."

Laura was the one most anxious to hear the answer. "And what did you answer?"

"Maybe."

"If you're not willing to give up school for the man, you probably aren't willing to die for him either."

"Mom, this time I'm not so sure."

"Megan, look around you. A ski resort is a virtual smorgasbord of feminine flesh. This guy probably has a new menu item every week. Please, don't be taken by a ski bum Casanova."

"Thanks, Brad. I think that was your attempt at caring."

Laura reached out putting her hand on her daughter's, "My Dear, when the right man comes along, at the right time and place, you'll know it. There won't be any doubt in your mind."

Krista and Chelli only heard the last line of the conversation as they approached the table. They were pulling on gloves expecting the others to be ready to face the challenging frigid air. The intensity of the discussion made it clear that something far too interesting was happening, and they'd missed it.

"What are you taking about?"

Megan didn't want to go through the explanation again. She didn't need to. She'd aired her frustration and felt better. The final decision would be up to her; she'd worry about it later.

Steel Illusions

"Thanks, Mom. Well, let's get going. The weather isn't getting any better." Megan reached for her gloves, pushed back her chair, and headed for the door.

"I think she feels better." Laura had forgotten Adam was sitting next to her until he commented about Megan. "Do you guys always have such open family discussions about sex."

"Is there a better place? Your day will come; before you know it, you'll be talking about boyfriends, cars and condoms."

"I think I'll just lock Susie up until she's thirty-five. Does Kevin talk to the girls about..."

"Sex and things? Sometimes. They usually tease each other more than talk serious; but yeah, he talks to them. And the last time I saw Susie, she had you wrapped around her finger. She can push your buttons; and when that doesn't work, she kisses you on the nose. Adam, you better be careful; you're a pushover when it comes to your daughter."

"You're right. I do have a hard time keeping a straight face when she starts laying on the charm. I can see it coming, and I still can't say 'no' to her."

"My hard work is done; yours is just beginning. I wouldn't trade places with you for anything."

"Speaking of places, this place is going to be snowed in before long. I think I'm going to have to take a rain check on the dinner tonight. If I don't get out of here soon, I'll miss my flight back to Dallas."

"At this rate, you'll be lucky to make the drive back to Denver. Are you sure you can't just enjoy the snow and stay over?"

Adam thought about the invitation before answering. He doubted it was really the invitation he wanted. "After last night, I think it's best if I try to make it back. I don't like cold showers any more than Megan does."

"Promise you'll be careful."

"Me? You know I'm always careful."

"Yeah, right. Come on. I'll ski down with you. In this snow, it'll take us forty-five minutes or so just to get you back to the gondola."

"Are you sure you don't mind going down early?"

"Not at all. I'm not tough enough to enjoy skiing in this weather. Krista loves to ski in the snow. She says it's the ultimate in fresh powder. I like a little less fresh powder and a little more sun. I'll tell Megan at the lift that I'm going back."

"Will you still go up to Beaver Creek for dinner?"

"Absolutely. Thursday night the kids ski down the mountain with torches."

"Real torches? Isn't it dangerous?"

"No. They're really using little glow sticks, but Krista and Megan always thought they sounded more grown up if they called it a torch. There's loud noise, which the kids assure me is something from the top forty list, and a wonderful display of fireworks. I went with them one year, but I didn't like skiing down in the dark without my poles. I have enough trouble staying upright when the sun's out. I usually wait for them with a warm drink in my hand."

"That sound's like my choice too."

"It's quite a spectacular sight to watch all those people ski down the mountain waving their lights."

"Maybe someday I'll bring Susie out here. It's about time for me to teach her how to ski."

CHAPTER FIVE

Kevin looked up and down the furnace walls. There had been no changes since his earlier visit with Dan, and he hadn't expected any. 9 East crane had been pushed out of the way and parts ordered for repairs. Clean-up crews were diligently moving through the shop, chipping away the steel and slag which had solidified everywhere. There were still no clues, no answers. OSHA wasn't commenting yet about their investigation.

Kevin knew, in his gut, that somehow bad scrap had gotten into the furnace; and his greatest fear seemed to be the only answer. At some point, closed containers had been dropped into one of the inventory piles. There was no way to track its origin which was the only way to find out who was guilty of shipping those lethal containers into the shop. Waterman was right. This was Kevin's responsibility. If he couldn't clean it up, he'd go down as one of the victims.

"Hey, Boss, some guy named Rick is on the phone. Said it was really important."

"On my way." Kevin headed for his office. Whatever Rick wanted to tell him didn't need to be subjected to the scrutiny of listening ears yet.

"What's up, Rick?"

"How ya holding up?"

"I'm trying not to think about myself; just get the work done. Did you talk to her?"

"No. I went to the hospital and talked with Mrs. Childers, Amy's mother. Amy had been given a mild sedative. Kevin, she's about three months pregnant. Her mother and doctor are concerned that the emotional trauma might cause a miscarriage."

"Oh wouldn't that be just terrific. Then she could blame me for the death of her unborn baby as well as her husband's."

Steel Illusions

"Kevin, as far as her husband goes, maybe it's better if he doesn't make it."

"What do you mean?"

"I've seen burn victims before. It can be pretty bad."

"I know. And Pete's burns were all over his upper body and face. He had no hair or skin left. I think Amy could handle it, but Pete has a lot of pride. He would have trouble being so terribly disfigured."

"Did you know Pete and Amy's daughter died a year ago."

"Yeah. It's been almost two years now. They took it pretty hard."

"Not only that; the hospital bills have nearly destroyed them."

"I didn't realize things were so bad."

"I did find out something else rather interesting from Mrs. Childers that I probably wouldn't have heard from Amy. She told me Amy complained all day yesterday about Pete's behavior. She was sure he was up to something dangerous, but he wouldn't tell her what it was."

"Are you insinuating Pete might have had something to do with the explosion?"

"It would certainly explain why he was down on the furnace floor without his coat on. Did you find his coat? Had it been damaged; unwearable? Did he forget it at home?"

"Oh, we found his coat; it was in his locker. I was hoping he'd be able to tell you why it was in his locker rather than on his body."

"Do you understand now why I'm hinting at the possibility?"

"I can't believe he'd plan his own death like this."

"Some men will do anything for family especially if a new baby's on the way. I have to break the story tonight, Kevin. If I don't, she'll call somebody else. The best I can do is offer an objective point of view. Tell me your side."

R. Z. Crompton

"Our standard practices put the safety of the men first. I have hounded them and hounded them about wearing cotton because it burns clean and to always wear the fire retardant coat. When someone is caught not adhering to the safety practices, he is disciplined with time off if necessary. Yesterday, Pete was the only man not wearing his coat, and he was the only one injured so badly. If you're asking me whether or not I think he is responsible for what happened, I don't want to believe he had anything to do with the actual explosion. However, I do feel he is responsible for his own injuries. Not only should he have had his coat on, there was no reason for him to be down on the furnace floor. His job was at the controls in the pulpit. Even OSHA hasn't faulted our safety standards. How's that for my side."

"It works for me. If I can't get in to see Amy, I'm going to tell her mother what to expect. I don't want to add to her emotional stress when she hears your side mixed in with her own. I'm sure she's expecting a one sided wash."

"Any other reporter would have been glad to run to the public with only her side of the story."

"Any other reporter would have expected you to return your phone calls. Remember, I had to track you down."

"So, I may have passed judgement too quickly. All those guys standing at the mill gate are going to be mad at you for getting this much out of me."

"Is there any more to the story?"

"Not yet."

"Then I gotta go. I'll let you know what happens. You may want to catch the news at five."

"Rick, thanks. I owe you."

"Not really. Getting this story makes us even. Besides, what are friends for?"

Steel Illusions

"Well, if you ever need any fence posts or rebar, I can get you a hell of a deal."

Friendly laughter filtered though the receiver, "Thanks, I'm sure my wife and kids will love steel fence posts and barbed wire around the house."

CHAPTER SIX

Laura was glad to have the afternoon to herself. Wrapped in a soft blanket, she curled up on the sofa with a good mystery and hot cup of tea. With a condo full of people, it was hard for any one of them to find some peace and quiet; so Laura was determined to cherish the little time she had. The crackling fire had chased the chill out of the room and out of her body. The wind gave an occasional howl reminding her of why she'd given up the slopes early. The forecast was for increasingly heavy snow into the evening hours. She wondered how Adam was doing on his drive back to Denver. She'd driven that stretch of I-70 many times. Vail Pass was the worst; at least, Loveland Pass had the Eisenhower Tunnel to offer some protection and a short respite from the elements. Begging mercy from Mother Nature and protection from God, she and the girls had spent many hours crawling along the interstate at less than twenty miles per hour.

Laura knew the kids were back long before they came through the door. A humorous din preceded their entrance. Laura heard an extra deep voice along with the laughs of her daughters. She assumed Luke and probably Stewart were along for the evening.

"Mom, where are you?" Megan yelled through the room.

"I'm right here. No need to yell."

"Sorry. Were you sleeping?"

"Not at all. I was enjoying the peace and quiet with a good book." Laura emphasized the "was" because the peace and quiet had been replaced with commotion and laughter. "You guys look like Eskimos."

Clumps of snow went flying everywhere as all six of them stamped their feet, shook scarves, and dropped gloves all around.

Steel Illusions

"How about something to drink and a snack? I already put a few things out on the counter."

"Thanks, Mom. Where's Adam?" Krista asked as she stuck a chip into her mouth.

"The weather was getting pretty bad, and he had to get back to town."

"I bet he's stuck on I-70." Brad hung his coat on a chair in the living room before sitting down next to Krista at the table.

"Why? Is it that bad already?"

"Yeah." Luke answered. "We heard there was a big accident just this side of the Eisenhower Tunnel. Traffic is backed up all the way down the mountain. Do you have a Coke? I'll have to save my beer drinking for after the night ski. We aren't supposed to smell like booze when we go to greet the kiddies." Chelli went to the fridge to get a coke for her and the others. She put the tray of crackers and cheese, Laura had prepared, on the table in front of them.

"By now, whoever didn't get threw the tunnel isn't going to. Vail Pass has been closed too."

"Mom, remember that time we got stuck in Dillon at the Pizza Hut? We were there for hours."

"I remember," Krista continued the story. "The whole town looked like a parking lot. What a long night."

"Adam is probably stuck in some restaurant for the night."

"Are we still going up to Beaver Creek?" Brad had no reservation about having a beer before heading outside again. The girls settled for hot chocolate laced with a little peppermint schnapps.

"Absolutely. It's more fun when it's snowing," Stewart offered.

"Do you mind if I just wait at the fire pit with something hot?" Laura asked.

"As long as what's hot isn't a man." Her oldest daughter smiled.

"I was thinking more about a drink, but you never know what you'll find up there. If I'm still alone when you're done, I'll take you all to dinner since Adam had to leave."

"Sure. That'd be great." Brad never lost the chance to agree about having a meal bought.

Luke was the one to disagree. "Since Megan's agreed to stay over for a couple of days, I'd like to invite all of you over to my place for dinner."

The silence in the room was colder than the snow outside. Everyone stared at Megan waiting for an explanation; but, as usual, it was Brad who broke the silence.

"So, Meg, you decided to follow your hormones."

Krista caught Brad's attention and whispered, "Shut up!"

"Megan, I'm really sorry. I thought you'd told them at lunch."

"I told them I was thinking about staying."

"Don't make such a big deal about it." Luke offered.

Megan was annoyed by his comment. "This is a big deal for me. Anytime a girl stays over with a man; it's a big deal, or, at least, it should be."

"Megan, 'girl' is the key word. When you become a woman, call me." Luke turned and headed for the door, but Megan was on his heels. She wasn't going to let the verbal slap go unanswered.

"You arrogant ass. Brad was right. It's guys like you who pass diseases to unsuspecting women."

"Listen, Miss Campus Co-ed, your reputation isn't any better than mine is. Do you think you're the only college kid on spring break? The girls come and go around here like flocks in migration, and they talk about their campus indiscretions. I had hoped you were different."

"Are you insinuating that I'm.....that I'm....." Megan was so frustrated she couldn't even finish the statement.

Steel Illusions

"Look, I didn't ask you to marry me just stay and ski for the weekend. My mistake, I thought it'd be a nice way for me to spend my one weekend off."

"Usually staying with a man for the weekend includes the indiscretion."

"If that's what you want, then send me a copy of a current AIDS test."

Before Megan could gather her steam to slap his face, Luke had closed the door behind him. Burning tears tumbled through her lashes and washed down her face. The irony was that the tears were not for the insult but because her heart was breaking for the loss; the loss of the man she thought she could love.

Laura understood her daughter's pain and humiliation. Walking over to Megan, she put an arm around her shoulder turning the girl away from the view of the others. Megan wiped the tears from her eyes and tried to get control of the sobs that were trying to escape.

Chelli graciously tried to lighten the mood. "So, I guess that mean's we're not eating at Luke's tonight."

Even Megan couldn't keep from smiling and appreciating the diversion. "He probably cooks like Brad. If he can't order out, we'd get peanut butter and jelly."

"Hey, I happen to enjoy peanut butter and jelly," Brad defended himself, "but not tonight. Any suggestions?"

"As the concerned mother, may I suggest we take the shuttle up to the village early and have a sandwich before you do your torch ski; then after the fireworks, we can come back here. I really hate to be out too late with the bad weather that's predicted."

Krista was the first to agree. "Actually, I wouldn't mind coming back here. It'd be nice to sit in the hottub for a while after freezing all day, and it's not going to be any warmer up there tonight."

"Mom," Megan spoke softly, "I think I'll just stay here."

"Nonsense, you need to go with us."

"That's right," Brad stated flatly.

"But Luke will be there. I don't want to see him."

"That's exactly why you should go. If we go and you stay here, Luke'll know he got the last word. Besides it's possible that he'll apologize."

"Fat chance! I shouldn't have hoped for so much from him. It's my own fault."

"C'mon, Meg, show some fight. You've never cowered before."

While everybody tried to build Megan's confidence, it was easy to forget that Stewart was sitting quietly on the sofa. He assumed a low profile would be best for the time being, especially since he was trying to convince Chelli to stay over for the long weekend. Most of his buddies owed him time off, and he finally had a good reason to collect. Stewart was silently hoping that Chelli wasn't taking all of this "disease" talk too seriously. He certainly hadn't been a boy scout over the years, and until Chelli showed up, he'd never worried about his sexual health.

"This afternoon you thought he was worth a fight."

"What can I do?" Megan looked at her mother for some words of wisdom.

"Well, Dear, my first suggestion is to go skiing tonight. See how you feel when you see him. If you still think he's worth fighting for, then send him a copy of an AIDS test."

Megan snickered at the advice. "You're joking. I don't need a test for that or any other S.T.D. I can't believe you'd even propose such a thing."

"Meg, you made some assumptions about Luke, but you don't know what kind of man he really is. How's he supposed to know what kind of person you are? He can't afford to take a chance any more than you can. Maybe he's different from other guys."

Steel Illusions

For some reason all eyes fell on Stewart. He felt branded with the word "scum" on his forehead. He fumbled with some words to defend himself and only thought about the picture they had painted of Luke.

"If I could interject some information here, it might help. I've known Luke for about three years now, and I really don't think he's a...." The words hung in the air, and Stewart hoped one of them would jump in and stop him from making a fool of himself; but everybody waited for him to continue. "Well, I've never seen Luke with any girl. He always goes home after a shift on the mountain or a rescue. When all the rest of us head for the bar, he goes home."

Stewart paused slightly to get his breath; and when no one jumped in to chastise him, he continued. "Luke's dependable and tough under pressure. We trust his judgement out in the back country. If I had to pick a single person to be lost in a blizzard with, it'd be him."

"Stewart, are you trying to make a point?"

"Yeah, I guess so." Stewart stood up and shoved his hands in his pockets. "What you're all thinking about Luke is wrong. Megan, your first impression of the man was more accurate than you know. A few years back Luke's parents were killed when their small plane crashed during a blizzard. Luke was away at school. Harvard or some fancy place like that. The story I heard was that he only had one semester to go, but he quit. He came home to take care of his two younger sisters and the family business. The older one is away at school now, but Stacy's still at home."

Brad wanted to test Stewart's defense of his friend, so he choose his words carefully. "If he's taking care of the family business, why does he have to do the ski patrol and the rescue?"

"The rescue work is volunteer, and we do it because it needs to be done and it's exciting. Ski patrol is part of his business. His family owns a good part of the investments that have been made in this valley. Being part of the ski patrol gives him the 'hands on' information he wants about people who spend their money in his town."

Megan was amazed at the revelation. If, and she still wasn't sure she believed what she was hearing, if this was true, then she owed Luke the apology. She hated apologizing to guys.

"You wanted someone who was different; well, he is." Stewart ended his defense with a helping of 'crow' for Megan and the others to swallow. "He's a father figure, business man, and volunteer. He doesn't' have time to party and carouse, and he certainly doesn't womanize."

"Ah, Stewart, I think you've made your point." Chelli decided he'd better shut up. "We all feel like idiots, and not all men are rotten. Now let's go eat, I'm starving."

"Stewart, do you still want to come with us, or will you be considered the traitor if you're seen with us?" Laura wasn't sure if he would be willing to forgive their transgression.

"Hey, up here, only a fool turns down a free meal. If I'm still invited, I'd love to come along. Megan, my advice, and I'm no expert on relationships, is to follow your mother's suggestion. You're the only woman I've ever seen get the man's attention. You cracked his wall of responsibility. He's finally thinking about something besides sisters and work, and I don't think he knows what to do."

"Thanks, Stew. We'll see what happens. Maybe I was simply the key to unlock the door, not the one to walk through it."

"You girls go freshen up, so we can get out of here." Before Laura could finish her command, the girls were out of the room.

"May I use the phone?" Stewart asked heading for the kitchen and leaving Laura and Brad alone in the living room.

Steel Illusions

"Laura, thanks for letting me come with you. I've had a wonderful time."

"No problem. You know you're like a member of the family."

"Thanks, I've enjoyed our relationship too."

"You sound like it's over."

"I'm off for New York this spring, so I won't be around as much."

"But we all assumed you and Krista would..."

"Get married?" Brad finished the half statement, half question for her.

"Something like that, yes."

"We are good friends. Maybe best friends. We date for awhile and then don't date. I'm merely her comfort zone until she gets done with school."

"Are you telling me she's using you?"

"Not at all. This friendship has been a mutual one. You have been more of a family to me than my own."

"But you've been dating since high school. And all the talk about love and looking at the 'scenery' on the mountain. I thought you two loved each other. I'd never have guessed you had a plutonic relationship."

"We do love each other, and I never said or implied we shared a plutonic relationship." Brad couldn't hide the grin on his face. "We have a compatible relationship which serves us both. We're each other's best excuse for not getting caught up in the campus party life. It's much easier to tell friends you can't go to a party because your significant other wouldn't like it than it is to say I'd rather study. Our good relationship kept us focused on classes and not entangled in distracting emotional ties."

"You make it sound so clinical."

"It's more complicated than I can explain. But I suspect someday a man will come along giving Krista a good dose of passion, and she'll get married over night."

"I never suspected anything like this from the two of you."

"Neither did we really. We'd talked about getting married several times, but our education and careers always got in the way. Even though we enjoy each other tremendously, if you know what I mean, that's not enough."

"When did you figure all this out?"

"I think we've always known, but never put it into words."

The conversation was interrupted when the girls came back into the room insisting that they were starving. Megan pulled her gloves on and headed for the door with the others close behind her. Laura turned off the lights and locked the door. They were too far from the door to hear the phone ring above the sound of the howling wind.

Megan didn't notice the chatter of the group as they piled into the shuttle bus. She was mentally chastising herself for being such an idiot. How could she have forgotten the words of wisdom she'd learned in school: assumption is the mother of all fuck-ups. Megan thought back to her eighth grade when she'd received her first and only "D" on her report card. She'd mistakenly assumed that the test day was the same for all classes: WRONG. Unfortunately, she never seemed to remember the lesson for more than a year or two before it was necessary for fate to do a rerun.

"Meg, why the pensive look?" Brad was the one to notice her far away look.

"Oh, I was just reflecting on the valuable lesson I learned."

"What's that?"

"Well, I guess I learned that not all men have a brain six inches long."

Steel Illusions

Brad tried to offer Stewart an explanation, "You have to forgive her. The whole family is slightly tilted. Sometimes their sense of humor is difficult to appreciate."

"I'm serious, Brad. Let's face it, that six inches is like a divining rod. You know that 'Y' shaped thing a water witch uses to find underground water. Where it leads: you follow - always faithful and obedient. It's refreshing to realize some guys, at least, try to coordinate some of their actions with the head on their shoulders."

"See what I mean, Stew?"

"Stop being so sensitive. You were the one who called Luke a ski bum Casanova."

"Yeah, I know; but then I was speaking more like a brother."

That remark stirred Krista, "So your 'coordination' is based on the role you're playing?"

Chelli piped in her opinion by adding, "Yeah, Brad. Who was the one who insisted on eating lunch at the condo yesterday - alone?"

"Guilty."

"Are you guilty too, Stew?"

"Wait a minute. I'm a new comer to this group."

"That's not an answer. Are you 'coordinated' or not. I would like to know." Chelli pushed for some kind of reply.

"I swear." He stood up, at attention with his hand over his heart. "I have been recently and totally reformed. I'm now double brained or double headed or whatever you want to call it."

Everyone laughed and the dark mood changed to a lighter side. Laura was glad; she wanted their last night in the mountains to be enjoyable.

CHAPTER SEVEN

"You guys go ahead and enjoy yourselves. I'm fine here.'

"Laura, are you sure? After last night, I really hate to leave you alone."

"What last night? What happened?"

"Some guy was following her. I saw him up here and at Cassidy's."

"So that's why you didn't want to go out with the others. Why didn't you say something?"

"He didn't need to. Adam got me back fine. There was no reason to worry. Besides, it was probably my imagination."

"Well it wasn't mine. And I'm not really crazy about this Adam either. He's too slick."

"Stop worrying. Do you see the culprit around now?"

"No, but in the snow and the dark, I wouldn't know if my own mother was walking past me."

"Go. Ski. What can possibly happen with all these people around? If I'm not here by the fire, I'll be in the restaurant staying warm."

****

"Meg, did you see Luke?"

"Yeah. A group of kids surrounded him before we got on the lift."

"Did he see you?"

"Would it matter? We leave tomorrow. I'm freezing."

"Me too, but it's still neat to be up here looking at all the lights. I love the night ski: it's Christmas and the Fourth of July all rolled into one."

Steel Illusions

"Tonight is more like Christmas in the Arctic. I'm ready to go down."

Everyone gave a resounding "me too".

As the lines of torches finally started the trek down the mountain, flakes of snow reflected the luminosity of the wand held by each skier creating a soft glow across the face of the mountain.

"Last man off the mountain is a flake, a snow flake." Brad quickly took off leading the group.

"You're the only flake we know, Brad." Megan yelled after him.

Lines of glow lights meandered down the slope. It was a short trip since no one had ski poles and the light from the wands wasn't enough to show rough spots in the snow. Music blared out across the valley abusing the beauty of the night. The five skiers didn't linger after reaching the bottom. The fire pit was their destination.

"Do you see Mom?"

"No, She's probably gone inside. Brad went to find her."

"Krista, she had the right idea, why don't we all go in. I'm freezing." Chelli was covered with a thin layer of snow just like the rest of them. She rubbed her hands together trying to relieve the burning sensation caused by the cold. Jack Frost was having a feast on exposed flesh, while a frigid north wind made talking difficult.

"You're right. There's no need for her to come out here when we all want to be inside."

Chelli didn't wait to see if any one followed her. She was desperately cold and ready for something stronger than hot chocolate to ease the chill she felt in her bones. She nearly knocked Brad over as he pushed through the door of the restaurant.

"She's not in there. I can't find her." There was a hint of panic in Brad's voice as he made his announcement to the small group.

"Please, Brad, let's talk inside," Chelli begged.

Brad didn't argue with her. He'd had his fill of Mother Nature for one day.

Krista could sense the concern in Brad's voice. "I'm sure there's nothing wrong. She's probably in the bathroom. I'll go check. Chelli, why don't you and Stewart order something warm for all of us to drink while we look for Mom."

"I'll go over to the hotel and check the restrooms there. She may be in the lobby waiting for us."

"That's a good idea, Meg. Ya know, she might be talking to Dad. Check the telephones too."

"Good idea. This wouldn't be the first time she called him while waiting for us. Remember when she called him from the top of the mountain?"

"Yeah, we waited fifteen minutes for her to show up. What I remember most is that you were sure she'd fallen off a cliff. Mom's number one rule has always been 'don't panic'; so, Brad stop worrying."

"I'm not panicking. I'm just concerned, especially after last night. I really did see a man watching her. I'll go with Megan. If there is some bad guy out there, I don't want her wandering around the hotel lobby alone."

Brad headed out the door with Megan while Chelli took off her scarf and unzipped her jumpsuit. She was ready for something hot to drink and looked around the room for a waitress. All the tables were occupied by cold skiers, some just wanting to warm up and others waiting for a meal. Buzzing conversation and simultaneous laughter surrounded the two left at the table. Chelli had assumed Stewart was going to join her, but he didn't sit down.

"Where are you going?"

"I don't like it."

"What? Having a drink while we wait?"

Steel Illusions

"No, having to wait for Laura is wrong. She was so concerned about getting back before the weather got bad that it doesn't seem right for her not to be here. I'm going to find Luke."

"Luke? Why? I don't think Megan'll like that."

"If they find her, it's no big deal. If not, then he's the one who'll get things rolling. He can use his radio to check with the other patrollers. Maybe somebody has seen her."

Krista was back within a matter of minutes without her mother. The concern could no longer be kept from her face. No matter how badly she wanted to think they just weren't looking in the right place, she a bad feeling in the pit of her stomach. Memories flashed through her mind of when she and Megan hadn't been waiting at the designated spot. The misery they must have caused their mother when they had gotten off the school bus at a different corner, or when they had gone home with a friend without calling Mom first. She wanted to scold her mother but didn't know if she'd do it before or after hugging her.

The sound of fireworks echoing through the night air caught Krista's attention. Searching for her mother was not the way she'd thought they'd be spending their last night at Beaver Creek Village. But then the weather wasn't exactly cooperating with them. They probably wouldn't be setting out by the fire pit watching the display anyway.

"Can't find her, Chel. Where's Stewart?" Krista inquired.

"He went for reinforcements."

"Huh? There's Megan and Brad, but no Mom."

"Krista, we looked everywhere. What about the condo? Could she have gone back?" Brad asked.

"Not a chance. She'd never leave without us or without leaving us a message where we could find it. Besides, I already thought about that and called. There was no answer and no message here with the hostess."

Megan was the first to notice Luke come through the door. Suddenly, she no longer noticed the chill in her bones. In spite of the fact she was worried about her mother, she couldn't deny the feelings the man stirred in her. The problem was that she didn't know if she was feeling passion or anger.

"What's he doing here?" Megan nodded her head toward the two men coming in the door. She decided anger was the safer emotion. It made her less vulnerable.

"Stewart was hoping he might be able to help us."

"I doubt it. If we wait here, she's bound to show up. C'mon, Guys, I really don't feel comfortable with him around."

"Relax, Meg. Stewart's right. Luke may be the fastest way to track her down so we can all go home."

Luke sauntered up to the table. Confidence oozed from him. This is what he did best: taking charge. No wonder the people working with and for him respected him; he wasn't arrogant about being in control of the situation. The roll fit him well. Only Megan felt slighted by the presence of the man. His confidence ate at her mask of indifference toward him.

"I assume you haven't found her since she's not here."

"Brilliant deduction," Megan chided.

Luke, ignoring her sarcasm, focused his attention on Krista and Brad. "I'll contact some of the other patrollers. I'm sure there's still a bunch of us up here. We can check out the obvious places up here faster than you can. If she still doesn't show up, I'll get a hold of Tim. He can start a search pattern that reaches out from here."

"Thanks, Luke, I...we appreciate your help," Krista offered. "What should we do?"

Steel Illusions

"Did you call the condo?" Krista shook her head. "Could she have gone somewhere with a friend? What about the guy from last night?"

"Which one?" Brad was the only one to think about the illusive man he'd seen staring at Laura.

"The one who took her home. Was there another?"

Krista offered the explanation. "She went home with Adam, but he left this afternoon."

"I doubt if he got far. The Eisenhower was closed by noon. By the time the accident was cleared away, the roads were closed because of the weather. Could he'a come back?"

"I suppose. But she wouldn't have left without telling us. We weren't gone that long."

"What about this other guy?"

"I saw him up here and down at Cassidy's. Laura saw him too. Then he disappeared."

"I remember. Mom looked terrified; and at the same instance, Adam showed up. None of us saw the other guy. Adam pretty much distracted us and Mom."

"I told you I didn't like Adam. I bet he has something to do with this."

"Brad, stop it. We've known Adam for years. He'd never kidnap Mom."

"Who said kidnap?" Chelli finally snapped to attention. She'd been quietly sipping on a steady stream of hot chocolate while the others played detective.

Luke quizzed Krista and Brad on obvious possibilities while negative reports came in one by one from the patrollers still on the mountain. They were lucky the weather had sent people home early and, yet most of the ski patrollers were still around. Once the skiing was over, only the restaurants and the hotels were open. Checking out

Beaver Creek was fairly easy. Had they been in Vail, it would've been impossible.

Luke's radio started to crackle. It was difficult to understand all of the words, but the general idea of the communication was clear. Somebody else was missing. Luke's demeanor didn't change; he asked smooth, direct questions as if he'd expected the problem.

"Does anybody remember seeing them come down from Grouse Mountain? They should've checked in hours ago"

The radio cracked again spitting back the negative response.

"Call Jackson's wife. Maybe she's heard from him."

"She called us. That's when we started checking with the patrollers around here. Things always get crazy with the night ski especially when the weather is bad. We hadn't even check the board yet to see who'd signed out."

"Okay, get a hold of Skally. I doubt if his crew is doing any grooming tonight. The wind should pretty much do the job for him. He can get a couple of his guys to run the big snow cats up to Grouse. He's got a lot better lights on those huge grooming machines than we do, besides the snowmobiles won't do us much good in this weather. Start a search with the cats and get a few patrollers up there to ride along. The more eyes we have looking for those two the faster we can all go home."

"What about the others? Several of the patrollers are ready to get out of here."

"Sorry, nobody leaves until we're all accounted for; so let's get out there and find them."

Megan's anger was starting to fire. She resented the way this "hero" had forgotten about her mother when his men were in trouble. After all, her mother was the guest; he should put her concerns above his own. Words of admonishment were forming in her mind when Luke gave his final instructions.

Steel Illusions

"That includes the missing lady too. Keep your eyes open. It's hard to imagine the two incidents are related, but you never know."

Megan felt like she'd been hit in the chest, but it was only her pride that was being deflated. She wanted to be angry at him. Unfortunately, she had no real reason. She had insulted him first; she owed him the apology. He was looking for her mother; she was being more of a hindrance than a help. She was lost in her own emotions rather than worrying about her mother.

"What can we do to help, Luke?"

"Stew, you go call Vail 1. With the weather so bad, I don't want to wait. I have a funny feeling we're going to need as many searchers as we can get. Laura isn't in the village, and I doubt if she'd gone up on the mountain willingly. Our first concern now is finding our own patrollers. The Vail Mountain Rescue will help to find Laura."

"Is that Tim?" Chelli asked.

"Yeah." Luke's short answer was interrupted by Krista's question.

"I don't understand why you can't do both."

"First of all, as ski patrol, we're only responsible for the ski area. When the area beyond that boundary has to be searched, we call in the Vail Mountain Rescue. Tim's the man who takes charge in just about any type of rescue mission you could imagine. He's the expert. Whenever someone is missing, we start with the immediate area and work out."

"Do you often have trouble with patrollers not checking in?"

"Rarely. That's why I'm concerned. It's dangerous up there after dark, and everybody knows it."

"Great. This is GREAT! Mom's lost in the dark, and we're sitting around a nice warm table talking about how dangerous the darkness is. What can we do?"

"At this moment nothing except stay calm. We only want to be searching for her: not all of you. Brad can you give us a description of the man you saw?"

"Not much of one. He was fairly tall, over six foot. It was hard to get a good perspective since I was sitting. He was dark skinned, not black; but very dark. His hair was coal black and slick. It almost seemed glossy in the sun. What I remember most was that he was dressed wrong."

"What do you mean?"

"He wasn't a skier. He had on city clothes and regular shoes, not boots like everybody else. No hat or gloves. He wore a black leather jacket and black dress pants. See what I mean? I didn't get the feeling he was on vacation. He didn't fit in. I saw the same guy at Cassidy's. He didn't fit in there either."

"Did Laura recognize him?"

"I don't think so."

"That helps more than you realize. As long as we know he's not comfortable in the backcountry, we can concentrate on the roads. In this weather, they can't get too far. All the roads are closed."

"Krista, we better call Dad."

"Not yet. There's nothing he can do but worry. What're we going to tell him anyway? Mom's been missing for an hour."

"I agree, Meg." Luke's comment grated on her nerves. Krista was right though. There was nothing their father could do. Finding their mother was up to them.

"My best suggestion to all of you for now is to get your gloves and shoes dry. Have something to eat and drink, but no booze. If we have to start a search, it's going to be a long night. You might as well be comfortable."

Stewart's great stride carried him across the room in just a couple of steps. Brushing the snow off his arms and shoulders, he announced, "Tim'll be here shortly. People aren't lingering outside tonight. Frankly, I don't blame them. Any word from the cats?"

"Not yet. They'll radio in as soon as they check out the top of the lift. Jackson and Sperry were supposed to close up the lift and head back in. That's where the search will start."

"Ya know, they might have gone over to the west storage shed to make sure everything was closed up for the bad weather. Do ya want me to take a ride out there? It's hard for the cats to get over there."

"Good idea, but take somebody with you."

"I'll go." Everybody was surprised at Chelli's offer.

"I don't think so."

"Why not? If I ride on the back of your snowmobile, I can be a second set of eyes for you. It'll be hard enough for you to see without trying to look around for somebody in the snow."

"I'll go too." Brad threw in quickly, "If you have another machine."

"Can you handle one?"

"Absolutely. This isn't my first trip into the mountains."

"I don't know. Stew, what do you think?"

"Frankly, I agree. Another set of eyes is exactly what we need. Besides the extra body warmth helps too. In fact, I think Krista should ride with Brad."

Luke didn't like letting guests ride in a search and rescue. It was dangerous especially in the bad weather, but he knew the more eyes looking the faster they'd find something. He still hadn't made a decision until Skally reported back that there was nothing at the top of Grouse Mountain Lift. Everything was closed up tight.

"Okay, go." He gave the order. "I'll wait here for Tim. Don't be dumb heroes though. Ride out, see what you find and get back here. If the cats don't turn anything up by then, we'll have to coordinate a full search. Tim should be ready to take over by then. We'll wait here just in case Laura shows up.

Megan sat alone at the table fumbling with the fingers of her gloves. She would've rather been any place but alone with Luke. She owed him the apology; and if she was going to give it, at least they had a moment of privacy. Better now than in front of everybody.

"Luke, I'm sorry about what I said this afternoon."

"Don't worry. It's over."

He brushed her comment off with no reservation, without even looking up from his pad of paper. To him, she was no big deal. At least she didn't have to feel guilty about hurting his feelings. She'd be back at school in a few days laughing about this whole affair. She tapped her fingers on the table to a steady rhythm of "I hate him. I hate him".

"Really, Meg, it's hard for me to concentrate with your pounding on the table."

"Fine. I'll go stand by the door."

Luke watched her stomp away from him. He hadn't meant to make her angry, but her feelings weren't his first priority. He was trying to make a check list of all the places between here and Grouse Mountain where the guys might have held up if one of them was hurt. The unexpected happened on a regular basis up here. As soon as he thought he'd seen it all, something new grabbed him. This wasn't the time to let his hormones start doing the thinking.

Steel Illusions

****

"Krista, have you ever driven one of these machines before?"
"Yeah, why?"
"Just wanted to know if you'd had any experience."
"I've had experience, but I don't know if you call it good or not. A couple of years ago we did a group thing up on the Continental Divide. I ran off the road and into a steel culvert. That was the end of my career as a snowmobiler."
"Well, I guess you won't be driving."
"What about you? I didn't know you'd even been on one."
"My experience is slightly better than yours: I didn't run off the road."
"Brad, what in the hell are we doing going along then. I thought you knew what you were doing."
"It's better than just sitting there waiting. Your mother was probably the most important adult influence in my life; I owe her. Besides, driving one of these things is easy. You just keep your eyes open. I'm going to have to focus on the path. Go get the suit and boots on. I'm going out to get a feel for the machine. And don't tell Stewart."

Krista waved him on. She wanted to go as much as he did. Something was wrong and sitting at a table waiting for news made her nerves raw. She mentally apologized to her sister for leaving her alone with the "man of her dreams". Krista was sure Megan was cursing all of them for abandoning her.

"Chel, are you sure you're up to this? It's going to be freezing."
Chelli didn't turn around. "Chelli?" "CHELLI!"
"What?" Chelli finally turned around, but still couldn't hear most of what Krista was saying to her. The helmet was hard to get on and off, so she wasn't going to wiggle out of it again.

"Are you sure you want to go?"

Chelli read Krista's lips more than she heard the words and simply shook her head up and down.

"Hey, Krista, where's Brad?" Stewart asked as he walked up behind her.

"He wanted to get the feel for the machine, so he went for a test drive. Why?"

"It's going to be really hard to hear anything especially with the wind blowing. Here's some headphones for you two. It's the only way we'll be able to communicate. But, just a word of advice, don't whisper anything you don't want us to hear."

"I understand." She smiled at him. "Hey, Stew, thanks for the snow gear."

"No problem, you'd freeze without it. You ready?"

"Yeah, let's get going. You fix up Chel's head set and I'll get Brad. Meet you two outside."

The smell of gasoline permeated the cold night air, but she was glad to get outside. The heat in the ski patrol office was causing her to sweat. Actually, she wasn't sure if it was the heat or fear, but she needed to get outside. What they were doing was dangerous. She momentarily relived the fear of her only snowmobile ride. Driving had been easy; steering was difficult. It required a lot of upper body strength, and she had no doubt that Brad was better prepared for the challenge than she had been years ago.

"Here's a headset for you, so we'll be able to talk with each other and with Stewart and Chel. They'll be out in a minute. Are you ready?"

"Yeah. It's difficult to see in the snow. The wind is blowing so hard it nearly whites everything out."

"By the way, don't mention your lack of experience with the head set on. They'll be able to hear everything."

Steel Illusions

"Get on. We'll give it a try."

Krista mounted the vibrating machine and took a firm grip of the handles on each side. The headlight didn't do much for visibility. Her knees squeezed tightly against Brad as the machine started to move. She remembered the strange sensation of being able to hear herself breathe. Whenever she inhaled it sounded like the hollow echo of a sea shell. She wouldn't be able to see where they were going. The best line of vision for her and Chelli would be to either side of the trail. Within minutes, she and Brad were following Stewart up the mountain away from the lights and warmth of the village into the darkness.

"How long before we get to the storage shed."

Instantly Krista heard a deep, masculine voice in her ear. "About ten minutes, Krista. How you guys doin' back there?"

"We're fine." Krista answered him. She felt confident about Brad's ability to handle the machine. His upper body strength was more than adequate to deal with the temperamental vehicle as they traversed the fluctuating terrain.

"Good. Keep your eyes open. If you see anything unusual, speak up."

"It's hard to see anything at all." Brad commented into the small speaker.

After several minutes, through chattering teeth, Chelli pleaded, "Stewart, talk to us. Tell us what trails we're on or something."

"We're following the path of lift ten to the top of Grouse Mountain. Just beyond is the boundary closure area. A hundred yards or so up in the trees there's a storage shed where we keep extra equipment and snowmobiles."

Krista instinctively leaned into the turn as she asked Stewart, "Why can't the cats look up there?"

"There's no path for them. They'd have to get out and walk through the snow several feet deep. We can drive right up to the door."

Chelli was listening to their conversation when she caught a hint of something off to her left. "Hey, I saw something over there!"

"Where? Over where?"

Chel's first reaction was to point, but felt silly when she realized no one would see her. "On the left!"

"Where? Imagine the face of a clock; give me a number."

"About one o'clock."

Krista saw nothing but black. "Must'a been your imagination, Chel. I don't see anything."

"You're right. It was just a flash. I....there!"

"Got it. A light of some kind at ten o'clock." Then there was nothing but blackness.

"Stew, could it be them?"

"I suppose, but we're almost there, so let's go ahead and check the place out.

Brad didn't see the shed until they were right at the door. It was a good thing Stewart new his way around otherwise Brad would have driven straight through the wall. Glad to get both of his feet on solid ground again, Brad could still feel the vibration of the snowmobile in his fingers.

Surprised to find the door unlocked, Stewart pushed it open and stepped into the small cabin flipping on the light. The others were close behind him. "Well somebody's been here."

"Can you tell if it was them?" Brad was the first to question the remains of a meal on the shelf.

"I'm sure it wasn't. None of us would leave our garbage behind."

"Who else would've been up here?"

Steel Illusions

"Somebody who knows the country, but isn't supposed to be in here. The door should've been locked."

"Good God." Krista was standing in the corner of the chilly room with a small pink tube in her hand.

"What's that?

"This is the lipstick Mom carries in her pocket."

Chelli was slightly more objective. "It isn't necessarily hers. That's a fairly common brand."

"Chel, think about it; how many women who might have a reason to be in this cabin would drop this tube of lipstick?"

"Most of the women who'd be up here for any reason don't use lipstick. Maybe chapstick or something, but not bright pink," Stewart offered.

Brad was as sure as Krista about the ownership of the tube. "I don't think your mom dropped it. I think she left it so we'd know she was here."

"Let's not jump to conclusions; however, don't touch anything else. Krista, leave the tube where you found it. We might be tampering with evidence."

"How do we know if it's evidence. After all, we don't know for sure who the lipstick belongs to?" Chelli queried.

"Maybe she left fingerprints on it. That's why we need to leave it here. Let the police take care of it. They can compare the prints of something you know is your mother's to the ones on the tube."

"Hey, that's good, Stew. Did you learn that from ski patrol?"

"No way. We don't do crime shit. I watch T.V." That brought a smile to an otherwise uneasy situation. Stewart took a small black radio out of the case hooked to his belt. He reported to Luke what they'd found, and the next instructions were specific.

"You guys get out of there and don't touch anything else."

"Do you want us to take a quick look around the outside of the shed?"

"Yeah, grab a couple of the big lights from in the shed. Try to get an idea of foot prints, see if any machines are missing. With all this snow, it'll be impossible to track anything tomorrow. Then get back here."

"Are the cats back yet?"

"No. They're still out searching."

"Did they get to the top of the mountain?"

"No. Why?"

"I was just wondering. The girls thought they saw a light coming from that area."

"I'll see if anyone can swing over to take a look."

Stewart opened a wall closet and handed large flashlights to each of his three partners. They tested their light and headed for the door. A gust of wind greeted them as they entered the night armed with a light which penetrated the snow and darkness only a few feet, but it was better than nothing.

"Let's stick together, guys."

"Aren't we going to destroy any evidence that might be out here?"

"Believe me, in this weather, any evidence will be covered in a matter of minutes. Our best effort is to see what we can as soon as possible, before its gone."

The four lights, working in unison, gave a respectable view of the grounds. There was nothing to see but snow and trees. As the line of light rounded the corner of the shed, it was easy to detect the inconsistencies in the snow. Strange mounds in neat light rows caught their attention.

"Stewart, what's that?"

"The extra snowmobiles. They're covered with tarps. All I need to do is get a count." Looking over the mounds of snow, everything seemed to be in order.

It was Brad who noticed the slight depressions in the snow. "Stewart, this looks like a track from a snowmobile. It leads away from the others, but there's a machine here." Brad crossed the area with his light. "It looks like several machines have been moved recently. Any reason for somebody to be up here in this weather?"

"Not really, and these tracks aren't that old. I'd guess with the wind and snow as bad as it is, we only missed them by a few minutes."

"The lights Chel and I saw through the trees on our way up."

"Could be." Stewart responded. "Brad, help me uncover these. I don't understand why the rows are all in order and yet all the tracks." Stewart handed Chelli his light and Brad did the same with Krista. The girls held both lanterns steady while the guys lifted the tarp off the end machine. Nothing. Stewart pulled off his glove and reached in under the cover to touch the side of the slick black body. It was wet.

"This machine has been uncovered recently cause it's wet. They shouldn't get wet under the tarp."

"Ste...Ste...." Chelli stuttered on his name when her light gave full view of his hand. It was red.

"Jesus Christ!"

"Is it blood?" Chelli grimaced.

Brad's retort was quick, "It sure as hell isn't gasoline." He reached for the tarp pulling it completely off the snowmobile. The body had slipped down behind the machine and lay crumpled in the snow.

"Is he still alive?"

"How in the hell do I know. Krista, you're the doctor in training. You check."

Krista approached the body. She noticed the bluish tint to the skin, but felt for a pulse anyway. "Nothing. I think he's dead, but I don't know how unless we move him."

"We can't do anything for him now, better not touch anything...you know, evidence."

"T.V.?" Brad asked.

"Yup. Let's see if we can find Jackson."

"Do you think Jackson killed him?"

"No, I think Jackson's dead too." Stewart moved to check another machine, but found nothing. The girls held the lights while he and Brad systematically checked under each snow covered tarp. The snow and wind hindered the handling of the already heavy covers.

"Here, guys check this one. It looks like there's less snow on it than the others." Krista was right. The blowing snow hadn't completely covered one of the middle tarps.

Brad and Stewart carefully worked their way through the maze. They didn't have to watch for foot prints because Mother Nature had already covered anything that might have given a clue. Finding the end of the tarp, the two young men fumbled for a good grip, so they could pull the cover back without destroying any possible evidence. Within seconds another body was exposed.

Krista was not crazy about checking another corpse, so she didn't offer her assistance right away. Besides, Stewart seemed to be doing the job for her. He was bent down over the body mumbling something about a "dirty, rotten son-of-bitch".

"He's alive!" Instantly the others moved toward the body.

"Stew, let's get him inside. At least, we can try to warm him up while we wait for help."

"Be careful. I don't know where he's hurt or how bad." Brad reached for the feet while Stewart tried to maneuver the broad

Steel Illusions

shoulders clear of the tarp. They had to lift the long body over the snowmobiles.

Krista and Chelli ran for the shed. "Here, Chel, you hold the light for them, and I'll see if there's anything we can use inside."

After the guys put the limp body on the floor, Krista tried to find out how badly Jackson was hurt. His pupils were dilated, but his skin didn't have the deathly blue tint of cold like the other one. The breathing was shallow, but constant. There had to be a wound somewhere; but with the poor lighting and heavy clothing, she couldn't find anything.

Stewart immediately radioed for help. It'd be several minutes before a rescue team would be able to get to them. He knew the routine. Even though everything was ready, and it was just a matter of getting a couple of patrollers out the door; there wasn't a faster way to travel the difficult terrain than by snowmobile.

"Are there any blankets or coats kept in here? Anything we can use to keep him warm?"

"Not really. We store equipment up here, not survival gear. There's a first aid kit."

"I don't think there'd be anything in it that would help. We'll have to take turns cuddling up to him, so we can keep him warm."

"Ya know, being up close to that warm machine probably kept him from freezing to death before we found him," Brad offered to the others.

"Who knows. It's possible being in the snow slowed down his circulation enough, so he didn't bleed to death," Krista countered.

"Then why did the other guy die?" Chelli challenged.

Stewart didn't want to make her sound silly, and he didn't have to. Brad did the job for him. "Ahh, let me think. The bullet in his head probably had something to do with it."

"Oh, I forgot."

"I know if we hadn't checked under those tarps, Jackson would be dead for sure. It might've been days before anyone got back up here. Brad saved his life."

"Yeah, I agree with that. Good job, Dear, " Krista commented. "Now move up here close to this man, so we can keep him alive."

****

Luke met Tim and several other members of the rescue team at the door. They were covered with snow and didn't venture into the restaurant. It was obvious to Megan that they weren't staying. She watched intently as Luke filled them in.

"Megan, we're going to the office where we can use our radios and spread out the maps. I don't think the restaurant appreciates our business at this point. Do you want to come along?"

"Of course I do." She grabbed her coat and gloves as she answered. She was actually surprised he remembered she was in the room, but then she'd already decided she was going to follow him around all night until he found her mother. In spite of her discomfort with him, she was confident he was the best person to find her mom.

Megan followed the men out the door into the stormy night. She tried not to think about her mother being dragged through the wilderness in this weather. What a wicked night. She quickened her steps.

Luke held the door open for Megan to pass. He felt sorry for her. The level of anxiety was the worst for the one left waiting. "Megan, if I give you a couple of numbers to call, would you order some sandwiches and drinks for us?"

"Yeah, Luke, whatever you need. I'd like to help."

Steel Illusions

"When Stew and the others get back, they're going to be hungry and cold. We may have to go out again, so I want food and drinks here when they get back."

"If they go, can I go with them. Please, Luke, I can't sit here by myself all night."

"Yeah, we'll both go." He gave her a brief smile and a piece of paper with two numbers on it. She quickly removed her coat and gloves and sat down at the desk. The time would pass easier if she was busy.

Sandwiches arrived before Stewart and the others. Megan was impressed that the second call Luke instructed her to make was for extra pieces of warm clothing, which arrived shortly after the sandwiches.

"Why the clothes, Luke?"

"It's not unusual to work up quite a sweat in this kind of weather. Stewart and the others will be more comfortable if they can change, especially the layer next to their skin."

"I see. When do you think they'll get back?"

"It should be any time. The ambulance is waiting to take Jackson to the hospital. Unfortunately, he should be flown to Denver, but the weather's too bad. The Vail Medical Center will have to do for tonight. They're better equipped to deal with ski injuries than with bullet holes."

"What happens next?"

"First we'll get as much information from Stewart as we can. Skally and his crew didn't see a thing, so all we have to go on is what Stew and his group can tell us. Tim will probably organize a search team to go out from the storage shed."

"But that's still resort property. I thought you covered the search on resort grounds?"

"Murder and kidnapping are in a slightly different category than searching for a lost skier. And I'm sure whoever did this is trying to escape. By the time we can get out there, they'll be into the back country. We have a much better chance of finding them if we turn this over to the experts."

Megan sat and watched as more members of the rescue team showed up. Luke and Tim hovered over their maps, and the phones rang incessantly. She wanted to help, but didn't know what to do. Finally, the room became so congested, she decided to just stay out of the way until her sister came back. Her fascination focused on the congestion. Megan found herself absorbed with the intensity of the volunteers. And grateful. They would be searching for her mother.

"Megan, they're back." Luke announced. He met Stewart at the door. "C'mon and get something warm to drink. Is Jackson still alive?"

"Barely, he's on his way to Vail."

"Did he say anything?" The men talked as they walked toward the immense map on the wall.

"Nope, he was never conscious. Krista's sure the tube of lipstick belonged to her mother, and I agree with her. Laura's not the type of person to just disappear."

"I agree with Stew, Tim. I've met the lady too. I think we'd be smart to assume she was in that cabin."

"So what are we going to do?" Krista wasn't going to let Luke and Tim make plans without her knowledge.

"We?" Tim turned around to see four concerned faces intently staring at the map on the wall. "I don't think WEEE are going to do anything."

Luke made the introductions. Tim remembered the girls and Brad from the night before, but he didn't know the missing lady was

Steel Illusions

Krista and Megan's mother. At least Tim understood now why they were so intent upon going along.

Megan caught the wink in Luke's eye and offered a compromise. "Tim, we'd just like to hear your plans. That's all. It'll help the time to pass." When Krista started to protest, her sister gave the family nod which meant "shut-up".

Krista caught on right away, but Brad was slow on the up-take. "No way! We've already been out dealing with dead bodies. If anybody is going to look for Laura, I'm going along."

"We don't have time to worry about you. You can stay and listen, or I'll have you leave."

"Brad, don't be irrational; we'd have no idea where to go or what to do."

"Fine!" He wasn't happy about not doing anything, but she was right. He'd be lost as soon as the village lights were out of sight.

Attention was again turned to the men standing directly in front of the map. "Stewart, approximately, where were you when the girls saw the light?"

"The first time, we were about here." He pointed to a spot on the map. "The next time, when Krista saw it, we were about here. Almost to the shed."

"Does that mean they were headed back to the village?" Krista asked the experts.

Luke explained the line of vision to her. "Probably not. Watch." He put a finger on the map representing Stewart's group and another finger for the light. "Now if you were here and saw a light at one o'clock, it would place the light approximately here. Then a few minutes later this was the location of the two points. As you made your way up the terrain you crossed the crest of the mountain. They seemed to be changing directions, but in reality you guys were just further up the mountain. If they had been headed back to the village, you would have been able to see their lights better and more than twice."

"How do you know there was more than one machine?"

"We're expecting there are two machines because that's how many are missing from the stock at the storage shed. They took two and put Jackson's and Sperry's back in the pack."

"They probably figured we wouldn't even look under those tarps for the missing guys. If Brad hadn't seen the tracks, we'd never have known the machines had been switched."

"It was just chance you went up there at all, Stewart."

"Krista, any idea why someone would want to kidnap your mother?"

"None."

"How 'bout your dad. You should call him, maybe he can give us some idea. It'd help us know if they're keeping her for ransom or......or....."

"I don't like the sound of that 'or'. I'll call him right away." Krista headed for a phone and dialed the house, but there was no answer, only the machine. "Dad, call us as soon as possible." She left the number and hung up the phone.

Brad was looking at the map with Luke. "At least the roads are still closed. We know they can't escape that way."

"Brad, if their original plan was to escape via the interstate or some smaller road and the weather caught them by surprise, then their options are limited to the stocked huts along the peaks. If they had planned all along to drag her out into the wilderness, we could be looking for days."

"What huts?"

"The Tenth Mountain Division keeps these huts stocked and fully contained during the winter. Most cross-country skiers know the huts are there and available in case of an emergency. Anyone who

Steel Illusions

uses them takes care to report it and replace the provisions." Luke pointed to several small dots.

Krista was enjoying the speculation. "If you guys know those cabins are stocked then wouldn't the bad guy be silly to go there."

"Not if they were caught in bad weather. It might be the only way they could stay alive. Besides, as long as they don't stay in one hut very long, it'd be hard for us to find them in this snow. As soon as the weather clears, we can get the choppers out. If they're trying to go cross country, they'll be easy to spot from up there."

"So what's the problem?" Chelli chimed in.

"Well, first, we don't know what they're planning to do with Laura; so we don't know how much time we have. Second, if they planned on hiding out in the wilderness, they would've stocked their own cabin which will be like finding a grain of salt in a bowl of sugar."

"Wait a minute. There's no way the guy I saw watching Laura would know the first thing about finding huts in the wilderness."

"Don't kid yourself. Just because he was dressed like a city-slicker, doesn't mean he was. You may have seen just what he wanted you to see. There's guys out here who'll do anything for a buck."

Stewart finished the explanation. "They're called poachers, and they're as bad as anything you've got in the city."

"Tim! Tim! You gotta a call." The patroller at the radio yelled.

Tim heard the frantic call for "Vail 1" over the radio. His attention went immediately to the voice coming over the radio. "There's been an avalanche at the Narrows on Vail Pass. We got people buried, but don't know how many."

"Okay, let's make some plans. Stewart, see how many of your patrollers you can get back to volunteer at the Pass. Seconds count. Tell them to meet on the scene. Luke, we can't search the high country for Laura until the weather clears, but we can have the

Highway Patrol check all the roads just in case they doubled back. My concern is this area here near Lower Turquoise Lake. You told me once your Grandpa used to take you down here fishing, and you'd stay in a tiny well built cabin."

"Yeah, I remember. It's well hidden; right at the base of Grouse Mountain."

"Well, I've heard tales from some of the old timers that the poachers have found a place to hide out in this area. Do you think it could be your old cabin?"

"I suppose someone might have stumbled onto it, but there's no way down in there. We used horses in the summer and followed the Beaver Creek Drainage down to the bottom."

"But look at where the girls said they saw the lights. There's nothing else down there."

"You're right; however, that's a tough ride."

"Here," Tim pointed at the map, "there's a snow cat trail that follows Beaver Creek almost all the way to Beaver Lake. You could take snowmobiles this far, but then you'd have to snowshoe in. This way you might be able to gain some time on them."

"That's a hell of a trek, Tim. If we're snowshoeing in, then they have to also."

"They may not make it. I'd hate to have 'em kill Laura because she couldn't keep up with them."

"Krista, can your mother handle the terrain out there?"

"She's fairly athletic."

"What if," Brad speculated, "What if the guys who planned this knew she was in good shape. Maybe they planned all along to have her hike through the wilderness. The weather is just helping to give them the cover needed to get away."

"Good possibility. Only idiots wander off into the wilderness without making sure they can get out. If they get tucked into a shelter

Steel Illusions

out there before we find them, they could stay hidden for days especially with this weather to cover their tracks."

"How do you suppose they're going to get out?" Brad asked the question, and everyone looked at Tim for the answer.

"My best guess is they've hired a chopper to come in and get them when they're ready. Luke, I'll send teams out to cover the obvious, but I have to get to the avalanche sight. You're the only one who can find that cabin. Pick a team you trust. Your best chance is to get there before they'd be watching for you."

"I can't pick a team. I'd rather take volunteers."

"Here's one." Stewart stepped forward after finishing his call. "Tim, a couple dozen volunteers are on their way; and they'll grab anybody who can swing a shovel as they go."

"Great! I gotta get out of here." Vail 1, intent on his mission, picked up his coat and headed for the door.

Brad was the next one to speak up. Luke wasn't surprised. He'd expected Stewart and even Brad to stand forward after their earlier discovery tonight; he'd also expected the girls to speak-up, but they didn't say a word. What he didn't understand was that they'd already exchanged their views on the adventure. Megan knew it was no use for them to present their case in front of Tim. He'd refuse to let them go, and Luke would be in a bad situation to counter the decision. But once Tim and the others had taken off for the Narrows at Vail Pass, the three girls would be able help the guys see a feminine point of view. They bided their time as the plans were made.

"Tim, when you get things under control at the Pass, you may want to send somebody up to check at Gilman. They might try to avoid a road block, by hanging out there for a couple of days."

"Good idea. Stay in touch. If... or when you spot them, radio in. I can get a chopper up there in minutes once we know where to go, and the snow lets up. Good Luck." Tim walked out the door leaving six volunteers ready to suit up.

"What's Gilman?" Brad asked.

"It's a ghost town tucked up on the edge of the mountain above Minturn."

"You mean the place with all the old, run down, white buildings and the sign on the gate that says 'Town for Sale?" Chelli inquired.

"Yup. Wanna buy it?" Stewart smiled at her. "Ya know, it's a famous place."

"Yeah, right." Brad commented cynically.

"It is. One of the Steven Seagal action movies, Under Siege 2, I think, was just done out here in the mountains, and they used the old ghost town for part of the train sequence."

"C'mon guys. Let's get ready." Luke started handing suits and gear to Brad. He hadn't expected the girls to be in line waiting for theirs.

Megan met his gaze straight on. "You promised we'd BOTH go."

"That was before I knew what we'd be facing. It's too dangerous."

"You need us. Or, at least, you need our eyes. If Krista and Chelli hadn't been along the first time, you'd have no idea where to look."

"Really, Krista, please try to talk some sense into her." Brad requested.

"I don't think so, Dear. Let's talk about your valuable experience on snowmobiles." Brad couldn't counter, but Luke wasn't ready to give in yet.

"Ladies, you just aren't strong enough to hike that far in the snow."

Steel Illusions

Brad smiled at the remark even before Krista could respond. This was one argument the males didn't have a chance of winning. Past experience had been a good teacher, and in the past he'd usually lost verbal debates with these women.

"We're just as capable of making this trip as our mother. Besides, I'm the closest thing to a doctor you have."

Megan had to add her defense as she pulled on a snowsuit. "Wait a minute. You may be the people doctor, but I've had more hands on experience with broken bones and injuries."

Luke looked at Brad and Stewart as Krista and Chelli fitted their headset. "Gentlemen, I think we've been had. Here, each of you get a pack. Take the small beacon out and turn it on. Stick it in your pocket and make sure you zip it closed. Just in case we're caught in an avalanche, it'll be easier to find you."

"This is supposed to make us feel better?" Chelli observed.

"You can stay here," Luke offered.

"By the way, what we told you about swimming in an avalanche is true, and keep your mouth closed," Stewart reminded the novice members of the team.

"Oh, now I feel much better," Chelli smarted off.

"Do we get a windshield wiper for this face shield?" Brad asked as he looked over the helmet with the hard shield attached to it and the microphone fit snugly on the inside.

"Yeah, your fingers." Luke answered sarcastically.

"Is there anything else we need to know about survival?" Brad asked.

Stewart finished their instructions. "Yeah, don't be stupid and don't panic."

Megan looked at her sister as they simultaneously commented, "He sounds just like Mom."

"Did you get a hold of your dad, Krista?"

R. Z. Crompton

"No, I called his beeper, but there was no return. At this time of night, there's no one in the office. All I could do was leave another message on the answering machine."

The search and rescue party headed for the door ready to face the blowing snow, but unsure of what they'd find in the back country of the Holy Cross Wilderness.

CHAPTER EIGHT

Amy sat at her husband's bedside. She wanted to hold his hand, to stroke his face, but there was no place left unbandaged for her intimate caress. He hadn't been conscious all day, so she couldn't tell him how frightened she was or how much she loved him. She wanted him to know she was angry at him for getting hurt, and she was angry at Kevin Bradford for letting all this happen.

The doctors had tried to be delicate in their discussion about Pete's condition, but she understood the reality. Her husband was dying. Amy sat in the chair pulled up close to the bed: the same place she'd been for more than twenty-four hours except for the brief respite she'd had earlier that morning. Amy looked at the figure, dressed from head to toe in white gauze. Even his eyes had been bandaged. She whispered softly in his ear.

"Pete, I'm here. I know you can hear me somewhere in there. Please, know that I love you. I need you. Don't leave me. Please don't leave me. I can't do this without you. Our baby needs you."

Her fingers automatically reached up to wipe the flowing tears away. Desperately, she tried to control the sobs fighting to escape from her throat; so she could continue the one sided conversation. "Why? Why?"

The groan coming from the whiteness startled her. Pete's raspy voice was barely audible, "Amy?"

"Yes, I'm here." Standing up, Amy leaned as far as she could over the body of her husband so she could place a delicate kiss on his lips. It was the only way she could touch him. The only place not covered in white. Pete winced at the pain caused by the brief touch.

"I'm sorry. I didn't mean to hurt you."

"It's okay. Amy..."

Steel Illusions

"Shh. I know it hurts to talk. I just want you to know I love you."

"You have....have to know." Pete choked out the words. "There is..... a letter..." He groaned again with the pain of uttering each word.

"Save your strength. Please don't try to talk."

"Amy....." It was barely a sound. "Amy, I love you....for you and the baby." Each syllable took significant effort. "Use the money for the baby."

"Pete, don't leave me." Amy didn't try to hide her sobs. "I don't want any money. I want you."

"Amy," he whispered. "Amy, I....I can see Molly...she's there in the light, waiting for me..." As his voice trailed off so did the last breath.

Before Amy realized what had happened, nurses were pushing her out of the way. She stumbled over her own feet as she was moved toward the door where her mother was waiting. There were no words, only emptiness. She knew it was over. He didn't need anything now. His pain was gone. She was alone.

****

Kevin looked around the shop. He was pleased with the work being done. 9 East crane had been moved out of the way, and the damage wasn't as bad as Sam had expected. OSHA had given the clearance for #2 furnace to start up in the morning. It'd be a week or so before the men could get all the steel and slag cleaned up; but, at least, the meltshop could meet part of its production plan.

"Hey, Boss. Things are looking better."

"Yeah, the guys are working hard."

"Mosely just called in. Pete died a few minutes ago."

"I'm sorry, Dan, but not surprised. How are the guys taking the news?"

"They're angry."

"Angry? Why?"

"They're mad at him for doing something so stupid."

"Have any of them said anything? Given any thought as to why he wasn't wearing his coat?"

"A few of his buddies thought he might have forgotten his coat, but he could have borrowed one. When we found it in his locker, that idea was shot. A couple of the guys have suggested he might have done it on purpose."

"What do you think, Dan?"

"I sure as hell don't want to think that!"

"Me either, but desperate people can do some strange things. Did you know Amy's pregnant?"

"Yeah, but that shouldn't make him desperate. I know when that woman of mine was expectin' I got pretty crazy. The bigger she got the crazier she made me."

The thought of Betty, big and pregnant, brought a smile to Kevin's face. She never minced words, and he was certain Dan had had his hands full. "She's still a handful, Dan. I know Pete had some good friends out here. See if you can get one of them to check on Amy. See if she needs anything. I can't call her. She blames me and the company for this."

"She's fixen to sue, isn't she?"

"I expect she is."

"But she can't sue the company."

"Not directly, but she can sue the scrap dealers and any other suppliers her attorney might link to the case. Unfortunately, the settlement she is entitled to will be sucked up by legal fees."

Steel Illusions

"She should take the money and forget any law suit. The company didn't do anything wrong."

"I know. OSHA hasn't found any fault in our safety practices."

"Did you think they might, Boss? Everybody around here knows he'd better follow the rules."

"How 'bout the Fire Marshall, did he get the information he needed?"

"Yeah, he left a couple hours ago with some samples of scrap from the explosion. I don't understand what cigarettes, beer, and guns have to do with the explosions."

"What are you talking about."

"He was going to send that stuff to some place dealing with beer and guns; it was a bunch of letters."

"The Department of Alcohol, Tobacco, and Firearms?"

"Yeah, that's it. What does a big government office have to do with this explosion?"

"They can run more specific tests than the fire department here."

"Maybe he found something that can help."

"Hope so. Now you go home and get a good meal and some rest. I'm sure Betty'll be glad to see you. I'll stay out here and keep the clean up going."

"Thanks, Boss. You get some rest too. By the way, did you ever call Laura?"

"No, but she'll be home tomorrow. By then we should have most of the shop cleaned up, and I'll be able to get home to see her and Krista. Megan's flying straight back to school."

"You'll be better off when she gets home." Dan smiled at his superior as he turned to leave.

"Wait a minute. Why'd you say that?"

"You know why. You're like a lost puppy without Laura around."

"Get outta here, you old goat." Kevin watched the older man turn and walk away from him before concentrating on the shop again. He wanted to get out to those inventory piles before it got dark. He'd been trying to get those damn piles cleaned up ever since he'd taken over the shop.

Several thousand tons of scrap had been stockpiled over the years. The majority of it was never recycled because loads were constantly being added to the top of the pile. Kevin knew the practice of stockpiling was the weak link in his manufacturing process. Gradually, he'd been reducing the amount of scrap in the stockpile so that it'd be easier to control what went into the furnace. All of Sam's incoming scrap records were in order, but it was possible that some of the really old stuff had worked its way to the top. Kevin walked out the door and headed for the back lot. He had time before the sun set to see what was hiding out there.

CHAPTER NINE

Laura was freezing. She'd been handcuffed to the snowmobile which was now racing though the snow. The brief scene at the fire pit played itself over in her mind, but she didn't know what she could've done to avoid being forced off into the darkness. A man with a gun was suddenly standing behind her. He didn't have to say much. The gun seemed to make the instructions clear enough.

"If you don't move very quietly up the mountain and away from the people, your oldest daughter will never see the light of day."

Laura had no reason to doubt his word, but then she had no reason to believe him. Just the fact he'd threatened one of her daughters was enough to make her do anything. She'd been pushed more than led away from the crowd and up the mountain into the dark. Another man was waiting beside two snowmobiles. Before she could get a look at either of the men, she was handed a snowsuit. It was obvious she was supposed to put it on. Then she was blindfolded and a helmet was forced down on her head. Handwarmers were placed in her gloves before she was pushed to sit down and each wrist was handcuffed to the grips on the passenger part of the snowmobile.

She couldn't hear any exchange between the two men. Only the howling of the wind managed to penetrate the helmet and hood she'd been forced to wear. Pungent exhaust fumes permeated the air as the energy of the machine vibrated through her body. There was no sense of direction only the side to side motion of the machine as they navigated the terrain. She had no idea of why they wanted her, or where she was being taken. Fear was held at bay only by her constant mental reminder to not panic. Giving the advice to the girls was certainly easier than taking it.

Steel Illusions

****

"Luke, how can you find this place in such rotten weather?"

"Instinct, I guess." Were the words coming through the headsets to the other five companions. "When I was a kid, I made this trip almost everyday. It didn't matter what the weather was like."

"Why didn't you ever build a road out here?"

"There was no need. Horses and snowmobiles were cheaper and didn't require such civilized things as roads. Besides, as a kid, I could ride the horse and snowmobile; but I couldn't drive. Once you cross the wilderness line, snowmobiles aren't allowed, so I was better off with my horse or on foot. I didn't have to worry about crossing the boundary line."

Megan tried not to touch the man sitting in front of her, but his large body gave off the heat she was craving. At least she didn't have to hold on to him. The grips on the side of the machine gave her the support she needed without the contact. She'd had no choice about which snowmobile to get on. If she'd made any verbal protest, she'd have been left behind.

"How far do we go on snowmobile?" Brad asked. He wasn't exactly confident in his ability to navigate the machine in the blinding snow. He focused on the red tail light in front of him. It was the only way he had any idea of where he was going. Risking his own life to find Laura was an easy decision. However, Brad felt bringing the girls along just gave him more to worry about. His responsibility was not just trying to find Laura; it was also keeping Krista, Megan and Chelli safe. He knew Kevin would want him to keep his daughters safe. On the other hand, Kevin had lived with these women for a long time and he was completely aware of how stubborn they could be. Even though he wished the girls were back at the condo safe and sound, they would've tied him to the bed post before letting him go without them.

R. Z. Crompton

I guess it's better having them along, so I don't have to worry about them taking off on their own.

Krista tried to watch the side of the path, but it was so dark and the snow so heavy seeing anything was extremely difficult. She felt the tension in Brad's legs every time the incline of the terrain changed. She knew he felt apprehensive about riding in these conditions, but she appreciated his willingness to join the hunt for her mother.

All of them were counting on Luke's ability to lead them safely into the back country. Luke was counting on the element of surprise, and surprise depended on the assumption that Laura was being taken to a specific cabin. This trip might be for nothing. Maybe, suspicion rose in the pit of Krista's stomach, maybe this was simply Tim's way of getting the girls out of the way so he could do the search without interference. Her curiosity provoked the carefully worded question.

"Hey, Luke, do you really think those guys are crazy enough to be out in this weather?"

"Yeah, poachers are more familiar with the back country than anybody 'cause they hide out there. I haven't been to my grandfather's cabin in years. It's probably been used for a hideout several times because it isn't easily spotted from the air. Besides Tim and myself, not many people even know it exists."

"So what's the purpose of getting there tonight?" Krista continued.

"We'll only get part of the way tonight. We'll have to camp out."

"Wait a minute!" Chelli squawked. "You never said anything about camping out. You're kidding. Right?"

Stewart tried to tactfully explain the situation. "We'll be fine, Chel. Luke and I'll dig a hole in the snow, kinda like an igloo. You'll be nice and warm."

Steel Illusions

"Yeah, right." Her words hung heavy with doubt. "If you knew where they were going, why couldn't we just wait until tomorrow?"

"First," Luke explained, "if the weather doesn't break up, the chopper can't search, and then we'd be too far behind to catch them. Second, we don't know what they're planning to do with Laura. The faster we get there, the less time she has to be at their mercy."

As soon as Luke paused, Stewart picked up his train of thought, "Besides, once it's light, if there's any trail at all, we'll have to find it right away. The high winds will erase anything not covered by the snow."

"We don't need tracks if you know where the cabin is." Krista still wasn't convinced that this wasn't a trick to get her out of the way.

"Tracks always help. They can confirm our suspicions that they're going toward the cabin or if they've veered off in another direction. Tracks will also tell us if they are still in front of us. I'd hate to have them sneak up on us."

"The odds are in our favor that the cabin is their destination, but I could be wrong," Luke admitted. "Because of where Jackson and Sperry were found and where you girls saw the lights, the nearest destination would be that cabin."

"So what happened to the theory that the bad guy was a city slicker and not good in the back country?" Brad's voice boomed through the headset.

"Jesus, Brad. You don't have to yell. We're right here not a million miles away," Chelli scolded.

"Sorry, I keep thinking I have to yell for my voice to get over the sound of the engines. So, anyway," he continued in a much softer tone, "what about the man I saw?"

"I have no idea. Maybe he was dressed that way to fool you."

"Maybe he's following a guide just like we are," Krista added.

"We're certainly making a lot of assumptions." Megan finally entered the conversation.

"So you remember your lesson in making assumptions from this afternoon, Meg?" Everyone snickered at her reference.

"What lesson?" Luke had been the only one absent for the earlier discussion and didn't like being left out of the joke.

After an uncomfortable pause, Brad tried to offer an explanation, "Oh just something Meg learned...."

"Shut up, Brad."

Krista tried to fill in the gap and save her sister's pride by moving the conversation along. "How much further to the stopping point?"

"We'll follow Beaver Creek drainage as far as we can. Then we'll have to snowshoe the rest of the way. We'll camp at the end of the drainage area."

"What about the bad guys, Luke? Will they have to hike in too?"

"Yup. There's no way anybody can ski or ride down to that cabin. I know, I've tried; and spent six weeks in a cast after I broke my leg. Now I realize I was lucky I didn't break my neck."

"So even you can make a bad decision once-in-a-while?" The sarcastic tone of Megan's voice told Luke that she wasn't expecting an answer. The remark was more of an observation than a question. He needed to talk to her, but it was impossible with all the other ears tuned in to the same conversation. There'd be no privacy until they got back to the village. She had been correct in her assumption about him. His reaction was purely a defensive move: he was ashamed of himself and didn't want her to know.

Even though he'd planned on having Megan stay with his sister in the big house, he wouldn't have turned down the opportunity to satisfy his body's urges either. It'd been a long time; and, in spite of the danger, he'd a bought a box of condoms and enjoyed his time with her. Luke had prided himself on being responsible, different from most of the other men on the mountain. In reality, he found himself running after a woman and planning the conquest just like the ski bums he criticized. What was worse, he had been willing to turn Megan into the kind of woman he despised. He sensed her hesitation to stay with him; he should've been more sensitive. Now they were both uncomfortable; and pride, at least his, was going to get in the way of his true feelings for her.

Luke was glad to have a diversion for the evening because he certainly didn't want the chance to look at himself in the mirror. However, he would've preferred a diversion which didn't include having the woman so close. Luke found himself smiling at the irony of the whole situation: he WAS going to be spending the night with the woman of his choice .....and her three companions and a fellow rescue worker. They would be sitting in front of a fire but one made in a coffee can not the big hearth with the bear skin rug in front of it. *I deserve this torture. It's my punishment.*

Luke felt Megan's thighs tighten against him as they swerved to miss the large tree. He needed to concentrate on what was really happening not what he'd hoped for, or they'd all get hurt. "Sorry, I didn't see the tree until we were almost on top of it. You okay, Meg."

"Yeah, sure." The answer was simple. Megan was deep in her own thoughts about the man in front of her. *What a fine mess this vacation has turned out to be. I shouldn't have come. Maybe none of this would've happened if I'd just stayed at school. I'm the fool. I would've stayed with him.*

"Slow down, guys. We're almost there." Luke was slowing his pace and didn't want Brad running into the back of him. It took a lot of guts to ride in this weather, and Luke respected Brad's courage. He'd shown no hesitation in going even after being out the first time.

Krista was glad when the snowmobile finally came to a stop. Her whole body seemed numb as she lifted her leg over the seat and tried to get her balance. The wind nearly blew her over as a gust howled through the trees. They were going to try and sleep in this. Fat chance!

Luke started giving orders immediately. "Krista, under the seat of each machine is a cover." He turned on his flashlight and showed her where the small storage compartment was. "You and Chelli cover the snowmobiles. There's a strap and hook here." He bent down and hooked one corner of the plastic cover to the other. "Snow and wind are hard on the machines. This gives some protection."

"Stew, there should be a good drift about twenty feet from this tree. You and Brad start digging the hole for the shelter. Meg and I will unpack some of the gear. Then we'll relieve you."

Stewart pulled a small shovel out of his backpack and headed for the drift with Brad trudging behind him. Brad had his doubts about spending the night in a snow cave. *What in the hell possessed me to drag my ass out here. We'll all be frozen to death by morning.* Then he thought about poor Chelli. She wasn't exactly the outdoor type and sleeping in a snow cave would probably leave emotional and mental scars forever. The humor of watching this pampered city woman survive in the wilderness was enough to spur him on. However uncomfortable he might be, she'd be tens times more. If they lived, this would be a grand adventure to tell his own kids about.

Luke finally had Megan off where he could talk to her without everyone listening. He removed his headset before talking to her. "Megan, we will be more comfortable tonight if we share a partner's body heat. I'm pretty sure the lines of partnership for the others have been drawn. That leaves you and me."

"I can deal with it. The physical discomfort of freezing to death carries more weight than my pride. Just tell me what you need me to

Steel Illusions

do." There was no animosity or sarcasm in her voice but neither was there friendship.

Luke admired Megan's ability to put up the stoic wall of protection. She was like him in that respect. He'd learned years ago how to hide his emotions from those around him, especially his sisters. When his parents had tragically died five years ago, the world crashed down on all of them. Both of the girls had depended on him to hold life together for them. They cried on his shoulder, in his arms. He had no one. He was thrown back into the family business because there wasn't anyone else capable of holding it together. Both of the girls were in school and too young to work even if they could. Luke didn't expect either of his sisters to give up her education just so he could stay out of the valley. He was too old for a guardian, and they were too young to be left alone; so his father had changed the family will only six months before the accident making Luke the legal guardian for both of his sisters and the executor of family estate.

Thinking about his father brought a smile to Luke's face. Dad had been grooming him to take over the business for as long as he could remember. It must have been Luke's destiny and the fact that his father didn't trust anyone else to control the vast holdings the family had acquired over the years. There were no other living relatives in the valley, just a few scattered aunts in other parts of the country. Luke actually had been happy to train under his father. While the other boys were playing basketball and going to school dances, Luke was with his father, learning the ups and downs of the ski business. The good times were hiking in the back country with his grandfather. He'd planned to attend Harvard and return to the valley, but his grandfather's death had changed his love for the wilderness.

Luke's grandfather, his best friend, companion, and only true confidant, had died the year before he planned to leave for college. Feeling bitter and lonely, Luke had left the valley wanting to never return. He didn't care if he ever saw a snow capped mountain again. He had been back to the cabin only to spread his grandfather's cremated ashes around the cabin grounds. The land had become a sacred place in his mind where he went during the lonely hours.

In the rescue mode, Luke hadn't had a chance to reminisce; but now that they were digging a snow shelter in his favorite spot, the old feelings were coming back. *Stay focused.* Luke reminded himself. *Do the job and get out. Don't think about him.*

"C'mon, Meg, let's relieve Stew and Brad." Luke headed for the mound of snow beginning to pile up.

"Stew," Luke yelled above the howl of the wind. "Take a break. Meg and I will dig, so you and Brad can get a drink. There should be room for two of us to dig inside."

"Sure, Luke. I'll check on the girls too. If we dig in shifts we'll be warm and cozy real quick."

"Okay, Meg. Let's get to work. I'll push the snow out, you remove it from the entrance."

"Sure thing, Partner." She answered without thinking about the possible implications of her term of endearment.

Luke liked the sound of *partner*, he liked the sound of her voice and her willingness to pitch-in. He put his headset back on so it'd be easier to communicate with her and the rest of his team.

"Luke, why is the entrance slanted upwards?"

"Because it keeps the cold air out of the cave rather than letting it filter downward into our living space."

"Ya know, it's freezing out here. How are we going to keep from freezing to death?"

"It's really weird; but even though the temperature will be about forty below, in here the temp will level off at about thirty-two degrees. With our special gear and shared body heat, we'll be comfortable enough."

Steel Illusions

Megan had her doubts, but she kept her feelings to herself. Chelli was the one to vocalize her opinion. "Excuse me, but I do recall that thirty-two degrees is freezing. We're going to freeze to death in a snow cave. How convenient! We're digging our own tomb!"

"Stop it, Chel. You're exaggerating. Do you think Luke and Stewart would've brought us out here if they didn't know what they were doing?"

"I suppose not."

Stewart was the one to finally put her mind at ease. "Don't worry, Chel. I've stayed in a snow cave lots of times. It's rather exciting. Besides, I'll keep you warm."

"Now there's an offer I like. Dig faster, you guys."

Stewart and Brad took their turn digging out the main part of the cave. "How big do we make this thing?"

"As big as we need it to be for the six of us to get some comfortable protection from the weather. We just have to make sure there's about a foot of snow for the roof and it's shaped like a dome."

"Why a dome?"

"A dome is pretty sturdy. A little bigger and we'll be done."

Luke pushed his microphone away from his mouth. "Megan, are you girls real modest?"

"Not really. We played competitive soccer for years. It wasn't unusual for us to change clothes in the parking lot. Why?"

"We are going to be sharing a very small space. You girls go in first. There's a dry set of long underwear for each of us in the backpacks. You change while we wait outside then we'll come in and change. Will that bother you?"

"Don't be silly, Luke. We'll trust you to close your eyes. There's no need for you to stay out in the cold."

"Are you sure?"

"Yeah, if Krista or Chel complain, I'll tell them they can wait outside."

"I need to put you in charge more often. C'mon, let's go. We'll get the big pack off the wagon Stewart was pulling. The sleeping bags are in there." The two of them carried the big pack to the opening of the snow cave where they began handing in the gear.

Within minutes a waterproof covering, filled with a thin layer of air for insulation, was placed on the floor. The girls quickly changed out of their damp clothing and replaced it with cold, but dry cotton underwear. Under the conditions there was no such thing as modesty.

As soon as the guys were dressed, Luke began digging in one of the packs. "There's rations in each of the packs. Let's have something to eat. I'll melt down some snow to drink."

"What. No wine for this romantic, winter picnic?"

"Sorry, Chel. Survival trips require no alcohol, but I have a stick of butter you can chew on."

"Yuck! You're kidding."

"How about some peanut butter or a hunk of salami?"

"I'd prefer something a little less caloric. How about some crackers?" Megan asked.

"Sorry, up here you really need to eat a lot of fat. Your body will burn nearly four thousand calories a day in these conditions," Stewart offered.

"That sounds great, but I've lived most of my life counting calories and fat grams."

"Meg, you always did try to diet." Krista acknowledged.

"That's because I've always had ten extra pounds."

"And you exaggerate. Stop putting yourself down. You can sing like a goddess." Chelli chastised.

"Oh, great! That makes me the fat lady who sings!" Everyone was caught off guard by her humor, and it took a moment to register before the laughter broke out. "Do you have a snickers bar in that bag?"

"You bet!" Luke tossed her a candy bar and a bag of peanuts. "Here, you can splurge. You earned it."

"Luke, are we going to sleep two in a bag?"

"I think that'll be the best. We'll be warmer and then we can use the extra bags as insulation. The little bit of air in this floor covering doesn't offer much protection. Put your shoes in these water proof bags, so they don't get the sleeping bag wet and then put them in the end of the bag. They'll be nice and warm when you put them on. It's terrible to put on cold boots."

"And keep your gloves and hats on," Stewart continued. "Ya know, I was wondering about your mother. These guys must have known she was in pretty good shape before they would've dragged her off into the wilderness."

"Yes, your point?"

"I get his point," Luke stated

"Me too," Brad added. "He think's these guys know your mother."

Krista was stunned. "Nobody would want to kidnap her."

"What about somebody who wanted to get at Dad?"

"That's a good possibility, Meg. You never did talk to him. Isn't that rather strange?" Brad's mind started backtracking, counting the incoming calls the past few days. Kevin hadn't been calling just once a day; he called frequently.

"I think I'll radio Tim and have him get in touch with somebody in Tulsa, who can track your dad down. If he can give us any ideas, it'd sure help."

"Ya know, it's weird we haven't been able to find him. What if he's been kidnapped too?"

"One worry at a time. Until you know something for sure about your dad, we'll concentrate on your mom. Now try to get some sleep."

Krista and Brad didn't waste any time crawling into their sleeping bag. She was looking forward to having the broad chest to snuggle up to and strong arms wrapped around her. If she could warm up, she'd be asleep in minutes.

Megan was the one to linger. It had been easy to pretend their relationship was all business when they were setting up camp, but getting into a sleeping bag with Luke was going to be difficult. She couldn't put it off much longer without his realizing her discomfort. After all, she promised her cooperation. Luke crawled in first and waited patiently for her. *It's business. Only business. Just go to sleep.* She reminded herself before accepting his invitation. As quickly as possible, Megan tried to get comfortable and stop moving.

Luke held his breath while the woman next to him tried to settle down. He was afraid if he made the slightest move he wouldn't be able to control the other parts of his body. The last thing he wanted was for this woman to suspect his true feelings for her. He chastised himself for even thinking about her sexually under the circumstances. *We're on a rescue mission for Christ's sake, not a romantic interlude.* Luke knew she was exhausted and patiently waited for her to relax and her breathing to slow before letting his arm slide around her. He wouldn't have been able to sleep at all in the rigid position he assumed. He knew by the relaxed breathing of the others, that Megan wasn't the only one sleeping.

"Stew," Luke whispered, "You asleep?"
"No. Why?"
"Just wondering how you and Chelli are doing."
"Nice and cozy. You?"
"Fine. Trying not to move."
"Ya know, you blew it this afternoon."

"Yeah, I know. Do you know what the lesson was Megan learned?"

"Yup." There was a hint of humor in his answer, but Stew didn't offer the explanation Luke wanted.

"Well?"

"I don't think she'd want you to know."

"Oh, tell him." Brad whispered through the dark. "He might as well know."

"I thought you were asleep."

"Not, yet, but Krista is."

"Chel, too. I explained to her that she was wrong about you."

"That's not exactly how Meg worded the lesson. She said, 'assumption is the mother of all fuck-ups' end quote." Brad finished.

There was no answer. "Luke," Brad continued.

"Yeah?"

"You were wrong about her, too. She's not...on campus, I mean....she doesn't..."

"I know, Brad. I was wrong about her, but I don't know if she was wrong about me and my intentions. Good night." Silence filled the snow, domed shelter.

CHAPTER TEN

Mrs. Childers sat quietly beside her daughter. Amy had fainted in her arms and was carried to a bed where she could be given a mild sedative again and get some rest. Pete was dead, so her main concern now was her daughter's well-being. She was worried about the girl. Amy had never stood alone, made her own way in life. She'd been in love with Pete since she was a little girl; adjusting to his death would be long and painful.

Finally the young woman stirred. "Mom?"

"Yes, Dear. I'm here."

"What am I going to do without him?" The tears began to flow freely down her pale cheeks.

"You're going to grieve and be angry, then pick yourself up and move on."

"I can't. I don't want to."

"Yes, you can. Pete would want you to take care of his baby. You have to put the welfare of the child first or you'll lose it. Do you understand?"

There was no answer, only a nod of her head. "Will you help me make the arrangements? I don't know what to do. I just want Kevin Bradford to pay for this. He's the one I'm angry at."

"Maybe you should be angry at Pete."

"What? How can you say that?"

"I was listening to the news tonight when the story you gave to the reporter was on. The man gave your side of the story and Kevin Bradford's. Frankly, he has a stronger case than you do."

"I can't believe it. Pete told me there was a letter. He always got night letters from Bradford with instructions for the next day."

"Where's the letter? If that's your only proof then you need to find it."

Steel Illusions

"I don't know. He died before he could tell me."

"Did he tell you why he didn't have his fireproof coat on?"

"He was dying, Mother." The sound of her voice was syruped with sarcasm. "He didn't exactly give me a run down of the chain of events."

A light knocking was heard at the door. "Come in." Amy answered.

"Excuse me, Ma'am." Amy wasn't sure if the nurse was talking to her or her mother. "There's a gentleman here to pay his respects."

Amy wasn't anxious to see anyone yet, but she liked the conversation she was having with her mother even less. "Tell him I'll be right there." She said to the nurse without acknowledging her mother's disapproving look.

****

"Ya know, Bradford was out there looking around." The deep voice bellowed through the receiver. "I sure as hell hope you had one of those drivers do a good clean-up job. I'd hate to have Bradford come across one of our special little gems."

"I'm not worried about a left over container as much as I am about the body that's hidden out there. It should be loaded up and put into the furnace tonight."

"Body! What in the hell are you talking about?"

"Relax, one of the drivers got too greedy; and, reluctantly, we had to change his mind. He won't be missed. What about Laura?"

"I'm not out there anymore. Remember? I had to come back before Bradford found out where I was. It's not like I can call my contact anytime I want. They are in the wilderness. Obviously, Bradford hasn't heard his wife is missing."

"He will. Then he'll be too worried to snoop around in scrap piles. Did you hear the kid died. Died without saying a word."

"How do you know?"

"I talked to the lady myself. Paid my respects so to speak. She's a very angry lady. I think we can count on her to add a few nails to Bradford's coffin.

"Is coffin figurative or literal?"

"Figurative for now, but I'm not giving up any options just yet."

"Did Pete talk to her? Did he tell her about the night letter?"

"I couldn't come right out and ask, now could I? She may be the grieving widow, but I doubt if she's an idiot. And it doesn't really matter as long as she thinks the night letter was from Bradford. She'll do anything she can to hang this around his neck. I promised to help her anyway I can, so I'll be able to keep an eye on what she's up to. Maybe there'll be an extra bonus for me. I mean if I get really close. She's a pretty little thing."

"You are so sick. How can you even think about screwing her? You just killed her husband."

"I didn't force him to leave his fuckin' coat in his locker. He killed himself. I'm just going to ease his wife's pain. I need to make sure Pete didn't leave any evidence around."

"What possible evidence. The only thing he could've left is the night letter with your prints on it."

"That's right. I'm going to make sure it isn't stuck away somewhere waiting to pique her curiosity. If need be, she may have to take her own life. You know: the grieving pregnant wife and all that shit."

"You wouldn't."

"If I have to, you can count on it. I'm starting to enjoy all this manipulation. In fact, finding Laura's body will do more for our cause than anything else."

Steel Illusions

"Wait a minute. Laura Bradford's my part of the plan."

"You fool. The kid's dead; that's murder. What's one more at this point."

"Pete was an accident. We didn't plan it. Her death is supposed to be an accident. She's supposed to just disappear."

"I'm sure the wilderness will take care of her for us. You just make sure your guys keep her...shall we say occupied for a while, and that she doesn't get away." That was the end of the conversation, but not the end of the thought. *I still think a body will have a greater impact.*

****

Kevin's inspection of the scrap yard last night just before sunset had stirred the suspicions he'd been trying to deny. The inventory piles had seen too much action the last few days. The scrap was delivered in railroad cars, unloaded into one of the various piles, and then loaded back into a gondola for transport to the furnace. So why the trucks coming and going? More questions and still no answers. He was sure of one thing: someone was lying to him. *Catch the liar; find the truth.* The shrill ring of the phone startled him.

"Hey, Kevin. I hear you're having some problems out there. What's up?"

"Adam. Where the hell have you been?"

"In some freaken snow storm in Colorado. I'm still caught up in a dive town in the mountains for at least another twenty-four hours. Shit there was even an avalanche last night. I'm lucky I can make any out going calls. Thank God for cellular."

"What were you doing out there anyway?"

"Had some business and then went skiing for a couple of days. I saw Laura and the girls. They were having a great time. What happened at the shop."

"Jesus, Adam! What a mess. Scrap blew up in the furnace. Killed a kid."

"Any of my stuff?"

"No, I don't think so. Probably some cylinders in the one of those old inventory piles out back."

"How long ya gonna be down?"

"We'll start-up #2 some time today, but #1 will be down for another week or so."

"Anything I can do to help?"

"Not really. Thanks for calling, but I've got a million things to do." Kevin hung up the phone. The conversation was flat, cold. He didn't have time for fluffy talk. He wasn't sure anymore who was telling the truth. His career was on the line, and his suspicions were raging.

As quickly as Kevin placed the receiver down, its ringing again filled the room.

"Hello, Kevin. It's Rick. I need to talk to you; can you meet me somewhere?"

"Yeah, but can't you come out here?"

"I'd rather not. Meet me as soon as possible at.....a....at Burger King."

"Okay. Be right there."

"Hey, don't tell anyone where you're going."

Kevin grabbed his coat and headed for the door. He didn't like the apprehension in Rick's voice. He nearly knocked Dan off his feet as he went around the corner.

"Mornin', Boss."

"Hey, Dan. You feelin' better?"

Steel Illusions

"Yeah. Did you get some rest? I bet you've been here all night."

"I'm on my way to grab some breakfast. Keep an eye on things for me."

"Sure thing."

A five minute drive brought Kevin into a busy parking lot. Rick was waiting for him at a table in the corner. "You sure got here fast, Rick. You got wings?"

"I was almost here when I called you. You want to get something to drink?"

"No, what's so wrong that you can't meet me at the plant?" Kevin asked as he slid into the seat.

"When was the last time you talked to Laura or one of the girls?"

"A couple of days. Why?" The single word was full of concern.

"This came in over the AP wire this morning. Some little paper in Vail put it out."

Kevin took the small piece of paper reading the words several times. "Is this true? Can this possibly be true?" He couldn't believe the words in front of him: *Laura Bradford was abducted from behind the Hyatt Hotel in Beaver Creek, CO last night during the night ski. Her two daughters joined the search and rescue team. Two victims, one dead and the other in critical condition, were found where the kidnap victim was held for a short time.*

"I called the paper in Vail, but there was no answer. So I decided to track down someone from the rescue team. They confirmed the story. Said the girls had tried to reach you last night."

"Yeah, I know. Another message I didn't take the time to listen to. Anything else?"

"The weather's impeding the search. Most of the search and rescue teams were involved in digging out avalanche victims. The girls radioed in a few hours ago requesting that someone track you down. So consider yourself tracked."

"What in the hell are my daughters doing out in the wilderness tracking down kidnappers?"

"They insisted on going with the others. I think stubbornness runs in the family."

"I suppose it does. They get that attitude from their mother, ya know."

"Yeah, sure. Kevin, they suspect the kidnappers know you and Laura."

"What? Who would want to abduct Laura?"

"Somebody who wants to distract you from other things."

"Shit!" Kevin ran his fingers threw his hair several times wondering who might be capable of planning such a thing. "That would mean the explosion was set."

"Yup. It also means the kid was murdered and you're being set up."

"What a fuckin' mess! I can't leave."

"Whoever set this up, I believe, is counting on the fact that you'll take off as soon as you hear the news. You'd be playing right into the plan."

"What if they plan to kill Laura if I don't go?"

"May I make a suggestion?"

"Might as well. I need all the help I can get."

"Flights to Colorado Springs and Denver are frequent, make several reservations for this afternoon, tonight, and even tomorrow. That way you can get out of here at a moment's notice; and whoever wants you gone, won't know when you're leaving. Have the rescue team radio the girls so they know what's happening here. Be straight with them. Then give the rescue team a chance to find Laura. You

Steel Illusions

couldn't do anything out there but wait. Here's a number, so you can call the rescue office direct, and I wouldn't do it from the shop."

"Thanks for the objective point of view."

"It's my job. And you have to find out who is setting you up and why. Any ideas?"

"I'm a steelmaker not an investigator. I have no idea where to start."

"I think you do. You're sure bad scrap was the cause of the explosion, so you start there. Who controls what comes in? Who has the most to gain by your absence? I'd watch for anybody who's anxious about your not going to Colorado to help search for Laura. And don't leave any evidence in your office or even make it common knowledge that you have it."

"The one thing that still doesn't make any sense to me is Pete. If he planned the explosion, then why was he down on the floor? You'd think he would've wanted to be as far away as possible."

"I agree, but what if he was only part of the plan. What if somebody wanted him there so he would get hurt?"

"It doesn't explain why he wasn't wearing his coat. You don't understand how significant that is."

"He might have had his own plan."

"Shit, this is getting more complicated every second."

"Crime usually is. At least, solving it is. Your thinking is too straight forward, Kevin. You're assuming everybody at the mill shares your ethics. Life doesn't work that way. Money motivates people to do some very strange things."

"I guess. My God! I talked to Adam this morning."

"Who's that?"

"A friend, at least I thought he was a friend."

"You have no friends right now, except for me, of course."

Kevin smiled, "Of course."

R. Z. Crompton

"I'm serious. The obvious isn't always the truth. Who's Adam."

"Laura and I've known him for years. He's one of my biggest scrap suppliers, and he was in Colorado the last couple of days. He called me this morning from some little town out there."

"Was it Minturn? The avalanche was near there. The A.P. wire I read this morning also mentioned suspicions about the avalanche being caused by human hands."

"He didn't say where he was, but I know Minturn is just east of where Laura and the girls were staying."

"Is he capable of violence? Murder?"

"Up until now, I would've said 'no'."

"I'll try to talk to Amy or her mother again, but I don't think she's going to be any good to us until she's had time to digest some reality. Her call to me last night was pretty ugly. For a lady, she has a very colorful use of the English language. Amy obviously didn't appreciate my objective reporting."

"I suppose not. Last night I was out by one of the inventory piles. There were several sets of tire tracks coming and going which shouldn't be there. That scrap always comes in by rail. I'm sure those tracks have something to do with all of this."

"Somebody stealing from you?"

"Looks like it. I think I'll go back this morning and take a closer look."

"Be sure to take several guys with you. It wouldn't be a good place to be caught alone."

"You're right. Want to come?"

"Really?"

"Why not? Bring a cameraman with you. Pictures will help me prove my point when I go in to see Waterman."

"I bet somebody will get real nervous when we go out there. Why don't you casually mention what we're doing and see who starts to squirm?"

"Sounds good. What time?"

"Give me three hours to follow-up on a couple of things. Don't forget to call the rescue office."

CHAPTER ELEVEN

Megan was extremely embarrassed to wake up wrapped in Luke's arms as if it was the most natural thing in the world. Her first conscious thoughts were about how comfortable she was. The secure warmth of the sleeping bag was amazing. As soon as she moved; however, memory slapped away the groggy, morning yawns. She was nose to nose with the man and his eyes were wide open watching her.

"Good morning. Did you sleep well."

"Considering the circumstances, very well. You?"

"Better than I expected. The others are still snoring. I'm amazed Krista can sleep at all with the noise Brad makes."

Megan didn't want to laugh, but she couldn't help herself. "Are you going to wake them up?"

"In a bit. We can't do much yet. It's still early. You can go back to sleep if you want. The day is going to be long and hard. We'll have to snowshoe the rest of the way to the cabin."

"How far is it?"

"We're about halfway. I won't have any idea how long it'll take us until I can get outside and see how the weather is."

"I'm surprised you can't hear the wind blowing in here."

"It's not too bad is it?"

"I wish it was under different circumstances. We might be able to enjoy the seclusion." *What a stupid thing to say*.

"I agree. Now try to get some more rest. It's going to be cold as soon as we open this bag."

Megan lay as still as possible, but there was no way she could go back to sleep, not with his arms wrapped around her. She could feel his breath on the side of her face and the constant rhythmic movement of his chest. It seemed like hours before she heard other movement in the cave.

"Brad," Krista whispered. "Are you awake? I need to get up."
"Why?" moaned a deep male voice.
"Don't be an idiot. Think about it. It's morning."
"Can't you wait until somebody else is up. I hate to wake everybody up."
"I'm awake, too." Megan whispered. "And I gotta go."
The girls crawled out of the warmth of the sleeping bags and headed outside to answer the call of nature. "Hurry up, Megan. It's freezing out here." Krista stamped her feet in an attempt to stay warm. "Why did we have to come so far away from the shelter?"
"Oh, Krista, humor me, would you please? I was so hot in the damn snow cave I thought I was going to suffocate."
"Yeah, we know why you're so hot!" Krista answered. "Let me tell you the cave isn't so warm. Your blood is boiling! You're making us suffer because you're so damn horny you can't think straight!"
Chelli came to Megan's defense. "Actually, I don't mind a little fresh air either."
"Oh, you're horny too, Chelli. Just hurry up!"
"Krista, what's wrong with you? I remember when you couldn't even wee outside. In fact you couldn't pee if somebody else was watching you."
"Yeah, and I remember you used to tell me just to close my eyes."
"It helped; didn't it? Eventually you peed."
"Well, you can't fix this as easily. I'm scared, okay. Mom's been kidnapped, and we can't find Dad. Aren't you just a little worried, or are you too wrapped up in your love life?"
"Yeah, I'm worried, but biting my head off won't help."
"Sorry, let's just get back. I want to get started."
When the girls crawled back into the cave, Luke was on the radio with Tim. "Okay, we'll be packing up after we eat. As soon as

you can find Kevin Bradford let us know. His daughters are anxious to hear from him."

Krista was sorry they'd missed the whole conversation. "Any news from down there?"

"Not much. Eat a big breakfast; you'll need the energy. The weather should start improving, but it won't clear off until tonight. Sorry, ladies, we may have to spend another night in the snow." Luke was surprised he didn't hear the loud protest he expected from the three women sitting on the floor of the cave. "I guess the good news is if we can't move very fast then neither can they."

Stewart had put a breakfast of dried fruit, nuts, and granola out for everyone. Milk and coffee were of the instant variety. A candle in the bottom of an old coffee can was the heat source used for melting down the snow. Stewart didn't say much as he played master chef for the group. He was pensive, almost furious in his actions. Finally Chelli asked him why. "I'd like to bury the son-of-a-bitch, who set off the avalanche, in a ton of snow."

"When did you find this out?" Chelli was mortified that somebody would have tried to kill all these innocent people on purpose.

"It came over the radio this morning. Five bodies have been pulled out so far, and Tim said they expect to find more."

"How does he know it was an intentional act?" Brad asked.

Luke offered the explanation. "Tim said some of the victims heard two loud explosions just before the avalanche hit. And the place is not considered a risk area."

"We heard an explosion the other day just before the snow started sliding down the mountain." Krista's comment was more of a question.

Steel Illusions

"That's just it. You heard one explosion. When we set off charges to start an avalanche, we hear two: One from the explosives and one from the snow breaking away."

Stewart took over when Luke stopped to get his breath. "When we use explosives, we might hear two quick pops; like firecrackers. But there isn't always a sound when the snow breaks away from the mountain. The loud booms are usually with the bigger slides."

"Why isn't the area high risk? It's at the base of the mountain." Megan queried.

"Mother Nature picks her favorite avalanche spots, and it's against the law to build in those slide areas. You can tell the runout paths; there're no trees. With the heavy snows all last week and yesterday accompanied by the high wind last night almost any moron can start an avalanche with a stick of dynamite."

Krista added a new dimension to the discussion, "What if it was done on purpose by the same group who kidnapped mom?"

"Why? I don't see a connection." Luke's face portrayed the furrowed lines of worry.

"Think about it. What happened as soon as Tim heard about the avalanche?"

"I get it." Chelli was excited, like she was answering a trick question.

"Me too." Stewart added before continuing, "Tim took all of the rescue workers, except for us, to the avalanche site because there were more lives at stake. The kidnappers would have time to get away in this weather without worrying about being tracked."

"Or," now Luke was thinking out loud, "maybe they didn't want to get away, just hide away. Some place where they wouldn't be found for a few days."

"You mean like a cabin hidden in the mountains?"

"Yup. Let's eat up and get moving."

R. Z. Crompton

The breakfast was absolutely delicious, not as good as Sunday Brunch at the Duck Club, but close enough. "Krista, when we get home, do you think Dad will take us to the Duck Club?"

"If he doesn't, I will. A Blackened Thick Tenderloin will taste soooo good."

"I prefer the Duck with Black Currant sauce, if you don't mind. And my own chair at the dessert table. And mother will have a bottle of the finest champagne." Megan added to Krista's eating fantasy.

Luke heard their exchange and wondered about the name. "What's the Duck Club?"

"It's our favorite place to eat dinner. The food is better than fabulous." Krista's voice oozed with desire.

"Yeah, and Alex treats us like we are queens. He's tall, handsome and has a luscious accent. He kisses my hand every time I go in there."

"Who the hell is Alex?" Luke didn't realize, as he said the words, how jealous he sounded, but he certainly got the attention of the others.

Megan answered his question with slightly exaggerated information. "He's the General Manager of the restaurant. Wasn't he voted the most eligible and lovable bachelor in town?"

"I think he was voted the best Manager not the most lovable man."

"No, I'm sure he's a bachelor."

Krista nearly choked before gaining enough composure to follow her sister's lead. "Oh, yeah."

"Wait a minute. I never heard that." Brad almost ruined the story.

"You were away at school. Besides you shouldn't be paying attention to eligible bachelors."

"I do if he's kissing my girlfriend."

"Relax, Brad, Alex has eyes for Megan."

"Are you dating the man?" Luke looked at Megan.

"No, I'm away at school, remember." She hoped Luke would take the answer, it was the truth, and drop the subject. She was sorry she'd toyed with his feelings.

"It's none of my business. I shouldn't have asked. Finish eating, so we can clean up the camp."

Megan was relieved when the crackle of Luke's radio changed everyone's focus. "Base to Team 1. Hey, Luke, you up there?"

Luke pulled his radio out. "Team 1 to base, we're here. What's up?"

"We just heard from a reporter in Tulsa, says he's a friend of Kevin Bradford. The Vail Daily must have sent out a report, cause the Associated Press picked it up. Guy in Tulsa just found it this morning, he wanted to confirm the story."

Krista blurted in, "Has he seen Dad?"

"Just a minute, I'll ask. Base, did you ask if he'd talked to Bradford?"

"Affirmative. Man said Bradford was fine and he'd get this information to him right away."

"Great. Thanks, base. When you see Tim again, tell him we think the avalanche was started by the same people who kidnapped Laura."

"Okay. But why?"

"As a diversion."

"Understood. Anything else."

"Nope. Out."

"Out. Hey, be careful."

"Thanks." Luke stuck the radio back in his pocket.

By the time Luke was done with the conversation, all of the food and paper had been packed away. Stewart had taken charge of the clean up. Snowshoes were handed out making room for everything that wasn't absolutely necessary to be packed into the small wagon on the back of Stewart's snowmobile.

"Chelli, get as much air as you can out of the mat. It has to fit into one of the backpacks. We may need it tonight. Krista, you and Megan pack a sleeping bag in the bottom of each pack and then distribute the weight of the other supplies evenly. We won't have any luxuries tonight."

"Luxuries?" Chelli questioned. "Did I miss something this morning?"

"You had food choices. Today, we eat what we can carry in our pockets. I think you'll prefer your sleeping bag to the smorgasbord you had for breakfast."

Holding a bright orange rope in his hand, Brad yelled, "Hey, Stew, do we need this?"

"Yeah, in case of a whiteout, we'll tie ourselves together so nobody gets lost."

"What about our helmets, Luke?" Megan asked.

"It's up to you, but your helmet will keep you a lot warmer than a hat. I'm going to take mine."

"Me, too." Stewart agreed. "Take your headset, so we can talk to each other. That way we won't have to use our energy by yelling or turning around. I'll make sure the snowmobiles are tucked in tight, and then I think we're ready to go."

Cleaning up the campsite had not taken long. The inclement conditions were not conducive to lingering for the sake of conversation. By seven o'clock the group was carefully trudging down the steep slope covered with several feet of snow. The rescuers traveled in a single file line with Luke in the lead and Stewart covering the back. The girls were in remarkably good physical condition, so they were able to travel faster than Luke had anticipated. Stops were frequent, but brief.

Steel Illusions

Luke insisted everybody have a drink of the melted water from one of the canteens.

"You have to drink, so you don't get dehydrated."

"I thought you only got dehydrated in the summertime," Chelli stated, waiting for an explanation.

"The air is so dry up here, you can become dehydrated without even realizing it. Once you're thirsty, the process has already started. Then you get tired and cranky. So drink."

"Luke, why'd you boil the water for the canteens this morning, why can't we just eat snow?"

"We have a little tiny protozoan up here that can cause some very unpleasant bathroom needs."

"No kidding." Stewart chimed in. "I was too lazy one time to boil my water first and did I ever pay for my stupidity. You should try getting out of one of these suits when you've got diarrhea."

"You've made your point, I'll drink only what you give me." Chelli gulped down a couple more swallows before handing the canteen back to Stewart.

"Before we take off," Luke got everybody's attention, "check to make sure your beacon, that little thing you turned on last night, make sure it's on and securely tucked back into your pocket."

"Why, Luke?"

"This is avalanche country at its best."

"Oh this is wonderful." Megan's words were full of doubt. "What's a little transmitter going to do for you if you're buried under a ton of snow?"

"It'll help those who aren't buried to find you. We'll be careful. This next area, we'll cross one at a time. See those trees over there?" He waited for all of them to answer affirmatively before continuing. "Good. I go first. When I make the safety of the trees, I'll let you know; and the next one can start over.

R. Z. Crompton

"Luke, why can't we go a different way?" Megan was absolutely terrified.

"It's not possible. I've been up here in the summer and looked for better ways to get down. There aren't any. Once you head out just keep moving slow and steady. Don't make any load noise, whisper if you have to talk." He didn't have time to baby her along. She'd have to move when it was her turn.

"Wait, why don't you make a whole bunch of loud noises and see what happens. Maybe you can cause the avalanche to fall before we cross."

"Nice try, Chelli, but it doesn't work that way." Stewart tried not to role his eyes, but he did catch the wink Luke gave him. It wasn't a dumb suggestion, just one from a city slicker who didn't know any better.

"Any noise we make could just loosen the snow so that the first person out gets caught. I know you're scared, but this is the best way."

"Okay, then let's get it over with." Brad was impatient. The longer they talked about it the more apprehensive he became.

"Megan, you come right after me."

"Yeah, sure, Luke," was her timid response.

Luke took the two steps necessary to look her straight in the eye. He moved the mouth piece away from his mouth so the others wouldn't hear him. "I'm serious." He tried to be sympathetic, but the words were harsh. "Don't let your fear put the others in danger. Stay focused on why we're out here."

Megan knew he was right. "I'll try. Really. Luke, be careful."

"I will. You too." Luke positioned his microphone again so he could communicate with the others and started across the white slope.

Megan could hardly breath as she watched Luke take each carefully placed step. He was a speck of black against the white. There was no land, no sky; only white. Megan knew he'd never see an avalanche coming; he'd just disappear in the white. Even the howling wind was against them. It disguised any warning sound the snow might give. Then it was her turn.

"Krista...."

"You can do this, Megan. Don't think about the fear. Mom is on the other side."

"I was going to say that you could have all of my clothes if I don't make it. Now I'm going to give them to charity."

"You can't give 'em to charity unless you make it, so get going."

Megan took her first steps onto the unprotected part of the slope. Tentatively, she made her way toward the sanctuary of the trees. Sweat, running into her eyes distracted her from the hesitation she felt with each motion. "Twenty-five, twenty-six," she whispered, trying to stay focused on the trees. She wanted to take her gloves off, and especially the confining helmet.

Krista recognized her sister's fidgeting as extreme anxiety. The constant pulling on her fingers and reaching to adjust her helmet. She had to help. "Meg," she whispered through the speaker, "think a song; think a song. Do Star Spangled Banner."

Luke and Stewart were dumbfounded. The last thing they needed was someone belting out the Star Spangled Banner. On the other hand, they didn't need her falling apart in no-man's land. Chelli and Brad knew what Krista's suggestion would do for her sister. Megan's music could sooth her soul, take away the jitters. Silence was all the others heard as Megan finished her walk across the white wasteland, but Megan's mind was relishing in the words of the national anthem. After crossing the magic line and entering the safety of the trees, Megan felt silly. She didn't want to face Luke or the others who seemed to cross without any trouble.

Everyone was silent until Stewart crossed the sanctuary of the trees. His words were the first they heard. "Krista, what a wonderful idea. That was the first time I've ever crossed a slide area without imagining Mother Nature standing at the top of the mountain whispering to me 'I'm going to get you, Stewart. You're not supposed to be here.'"

"I didn't know you had anxiety attacks when you crossed a slide prone area." Luke sounded honestly surprised.

"Not this time. I came all the way across thinking 'Happy Birthday to me. Happy Birthday to me.' It worked!"

"Was Happy Birthday the best you could do?" Brad teased.

"Yeah, it was the only song I knew all the words to. What about you? What did you sing or did you just piss down your leg?"

"No problem here. Twinkle, Twinkle Little Star has a nice swing to it. Don't you think?"

"Twinkle, Twinkle Little Star?" Krista taunted. Everybody laughed.

"Well, I started out with Rudolph, the Red-Nosed Reindeer but I couldn't remember all the words. It must 'a been the pressure."

Megan felt so much better knowing the others had been afraid also, but she didn't really believe them. "C'mon, you guys. I appreciate the fact you're trying to make me feel better, but Twinkle, Twinkle Little Star? I can't believe that."

"I do." Krista offered. "He can't sing anything. I'm surprised he even knew all the words to a song."

"Whatever works." Luke seemed pleased with the new technique. "Once I had to go halfway back and pick up a man who'd fainted because he'd been holding his breath. That scared me more than crossing the first time."

"Really, Luke? I can't imagine being so scared that I wouldn't breath," Megan laughed.

"It's pretty common. This is the biggest slide area we have to cross but it's not the only one, so let's get moving."

"Can we talk?" Megan asked.

"Sure. Why?"

"Cause I was wondering what the chances are of accidentally coming up on Mom and the kidnappers?"

"Until the snow stops, visibility's so bad we'd have to step on them in order to see them." Luke answered. "If the weather report was correct, the snow won't start letting up until this evening; we should be at the cabin long before then."

Stewart wasn't as concerned as the others. "Do you really think they kept moving all night?"

"Yeah, that's my guess. I'm sure their idea was to get to the cabin as soon as possible and then rest. I doubt if they wanted to be digging a snow shelter and setting up a camp," Luke answered. Then he pursued another thought. "Hey, Stew, do you really have visions of Mother Nature waiting to crash down on your head?"

"Absolutely, last time I was out I imagined a beautiful woman with long wavy blond hair, her white gown billowing in the wind. She was yelling, 'Stewy, you're playing chicken with me and you're the chicken.' I was sure I was going to die that day."

"Stewy?" Everybody snickered at the boyish nickname.

"Yeah, that's what my mother used to call me."

"Oh, so this is a mother thing?" Krista commented.

"That was her pet name for me, but she never buried me in snow."

"If you have such anxiety attacks, why do you volunteer for the rescue team?" Luke asked.

"I love the excitement," he answered.

Chelli was amazed. "It must be one of those stupid, macho things."

"No, he's just thinking with the wrong head." Megan's teasing brought giggles from the three girls. "You know, the six inch one!"

"Wait a minute, I resent that."

"Actually, Stewart, I think you resemble that remark. And you told me you were a reformed man." Chelli pouted.

"I am. Now I sing instead of think."

The group proceeded, shuffling the cumbersome snowshoes through several inches of new powder. Their pace was slow as they maneuvered through the uneven terrain. Fallen trees, boulders, and creek beds provided challenging obstacles. It was obvious why snowmobiles were not an option out here. The descent was incredible. Only by reaching from one tree to the next were the girls able to keep their balance. Megan gasped as her snowshoe slipped out from under her. Not wanting to slide down the mountain head first, she reached for a tree branch, but it snapped in her hand. Before she could lunge for another limb, Luke had her safely tucked in his arms.

"Thanks, I almost beat ya' all to the bottom."

"You okay?"

"Yup." Megan came up looking directly into Luke's eyes. She could feel her body temperature rise. He had a tenderness she liked, but an attitude about women that she didn't care for. Without being obvious, she tried to step away from his support. Nothing would be accomplished by arousing his interest again. She needed a new conversation to help get her mind off this man's body.

"Luke, I was wondering why you stopped going to the cabin? You know the land so well, you must've come up here a lot."

"I did up until my grandfather died. Then there didn't seem to be a reason any more." He almost stopped talking and then decided it was time to tell the story, maybe it would help him face the past when they got there. "When I was a kid, I came up here weekly. I had all kinds of reasons for coming up here. I usually came with my grandfather, but sometimes with a friend. There were even times when, against my mother's direct orders, I set out alone. I loved the solitude of the mountains and the beauty of the valleys. Favorite animals had names and waited for me to ride by. Sampson, my horse, could make this trip in less than a day even when we stopped for lunch."

"Speaking of lunch, when do we get to eat?" Chelli interrupted.

"We can have a quick break now. Is everybody ready?" The affirmative nod from the rest of the team meant it was time to stop.

"Luke, weren't you ever afraid of the wild animals?" Megan was interested in his story about the back country.

"What wild animals? Why didn't anybody tell me about wild animals?" Chelli started looking over her shoulder.

"Don't worry," Luke smiled, "there's a few black bears around, but they're all hibernating now. Once-in-a-while we get a mountain lion. That was more exciting than scary. I was never afraid of the animals. Man was a lot more dangerous."

Lunch consisted of granola bars, dried fruit and water; it was nutritious and filling. Consumption took only minutes, and they were back on the trail again. The challenge of navigating the ups and downs of the mountainside was draining them mentally as well as physically. Concentration was absolute as they struggled to keep the snowshoes from getting snagged in the tree limbs. Snow was so deep it covered the bottom branches making them difficult to see until stepped on. When the brush opened up slightly, Megan asked Luke to continue his story.

"So what happened to your grandfather?"

"We were up at the cabin. I was fishing for dinner, and Grandpa went to chop some wood for the fire. I don't know the details, no one does. The best I could figure was that a poacher snuck up on him. Maybe the man had wanted to use the cabin and hadn't expected anyone to be around. I heard the gun shot. My gut ached as I ran for the cabin. I knew. I knew it wasn't a shot from his gun.

I knew he was dead." Luke's voice cracked, and he stopped talking to gain his composure. None of the others said a word. Stewart had never heard this part of Luke's story.

"I saw him laying on the ground near the wood pile. He'd been shot in the chest. I found the ax covered with blood several feet away. I assumed, by the trail of blood leading into the forest, that Gramps had thrown the ax at whoever shot him. I put Gramp's body on my horse and walked him out. I brought him back after he was cremated. This was his place. It was where he belonged, not in some flowery park filled with headstones." Before his voice cracked, Luke stopped again.

Megan was the first to speak, "The man was never found was he?"

"No. How'd you know?"

"It's in your voice. I'm sorry."

"Thanks. It was a long time ago."

"I mean about making you come up here. This must be very difficult for you."

"It's time for me to face the past and put the pain to rest."

"Hey, Luke! Look at this." Stewart was standing next to a tree, just a plain tree. No one could see anything unusual about it.

"What?"

"Look at the way these branches are broken off. The old break here," he pointed to a branch, "it's weathered so it's the same stone gray as the rest of the tree. Look here," he pointed to another spot, "these breaks are fresh." Stewart reached up and snapped off a small twig revealing the same light tan color.

"You're right. Somebody's been this way." Luke agreed. The sadness in his voice was replaced with excitement.

"Maybe it was just the wind." Brad wasn't so easy to convince.

Steel Illusions

"I don't think so. The breaks go to this tree, then to this one." Luke sounded excited as he went from tree to tree showing them the new breaks. "They must've been by here a few hours ago 'cause there's no foot prints: too much snow and wind." Luke followed the breaks where the trees ended. I bet if we angle downhill, we'll be able to find their trail on the other side of the clearing.

"Hey, Luke, did we ever come up with a plan for when we get to the cabin?" Stewart questioned as the group hiked on.

"Tim told us to radio in, and he'd send help as soon as possible."

"I don't like that plan at all," Brad declared.

"Me either. We could be waiting outside that cabin while some man destroys my mother. There has to be something better." Krista was almost angry as she spoke, and her voice seemed to echo through the valley.

Luke offered the best suggestion for the moment, "Let's get there first. We'll worry about a plan after we find them."

The clearing wasn't very wide. The rescue team proceeded single file across the slope thankful for the break in the difficult terrain. Before Stewart had entered the clearing, Luke was already searching the opposite line of trees for the easiest entrance into the heavily wooded area. Only Luke and Stewart recognized the sensation of sliding snow.

"Shit! Hold on guys! Here we go!"

At first the others didn't understand, but only an instant was needed for full recognition. There was no time to communicate. The power of the snow pushed them off their feet immediately. Tumbling each, head first, down the side of the mountain.

Megan was terrified as she felt the snow pulling her under. *Swim! Swing your arms!* She screamed at herself. She tried to take big gulps of air; but when her mouth started to fill up with snow, she clamped it shut. As the snow pulled her down, she felt her arms and legs getting caught up in the frozen debris already collected by the force of the snow. There was no time for fear, or for panic; she had to think. It was extremely difficult to move her arms in the heavy snow, but she forced herself to make the swimming motions Luke had told her about. She tried to imagine she was in the warm waters of the tropics enjoying the rolling waves.

    The ride seemed to last forever, but it was only seconds before Megan had the sensation she was slowing down. It took every ounce of strength she could muster to pull her hand up in front of her mouth to create an airspace before the snow came to a halt. To her amazement, a couple more tumbles in the snow and she came to a complete stop in a sitting position with only her legs buried. There was no one else in sight.

CHAPTER TWELVE

When Rick arrived at the mill just after twelve noon, he waited for Kevin at the front office rather than seeking his own back entrance. Most of the other reporters had been satisfied with the official company statement which had been given the day before. Rick had laughed when he'd heard the report. His story, which aired just after the company's statement was issued, had substance, passion and objectivity. He was satisfied, for now, even if Amy felt her side of the events were slighted.

"Sorry, I'm so late." Rick offered as he came up to Kevin. "I tried to see Amy, but she's not speaking to this reporter. I had to track down her mother."

"What'd you find out?"

"A man stopped by the hospital last night to 'pay his respects.' Respect is the key word here. I think it might have been Waterman; and based on what you've told me, I doubt very much if there was any respect involved. I think it was curiosity. He was fishing. Speaking of your boss, how's he taking all this?"

"Before or after your story broke? He thought he'd issued a generic statement which would pacify everyone. Then you gave, in my opinion, a wonderful report; and the phones started ringing like crazy. Shit, I thought he was going to rip me a new asshole. He accused me of undermining the company. I said I was just trying to find the truth, but he didn't care for my side."

"Does your boss or any ever really care about the other side?"

"I suppose not. He's not even offering to help solve this mess. He's waiting for me to fail, so he can fire my ass."

"Maybe you should ask yourself what he has to gain by your absence. Did he say anything about your not going to Colorado?"

"No. He didn't say anything at all."

"Nothing?"

"He asked how Laura was, and when I'd heard from her last."

"Is it normal for him to ask?"

"Come to think of it, no. He rarely asks about Laura or the girls."

"What'd you tell him?"

"I just told him there was a little problem, but the girls were taking care of it. I can't believe I'm not on the first flight out of town. I should be. Laura and the girls are more important to me than this job. You know that, don't you?"

"Yeah, I know. In fact, I'm surprised by your self control. Frankly, I don't know if I could handle things as well as you are. My gut instinct would have me on the next flight and then trudging through every drift of snow until I found her."

"It's nice for you to be so honest with me since it was your suggestion that kept me here."

"I still think you're doing the right thing. It isn't going to do Laura any good to have you lost in the wilderness. I'm sure she'd rather have you clean this mess up. I know we're still guessing about her kidnapping being a diversion to get you out of town. The last thing you want to do is what the puppet master expects."

"I hope you're right."

"Of course, I am. I'm your anchorman." Rick laughed. "Get it?"

"Yeah, I get it. You keep my feet anchored. Thanks. Now I can see why you don't write your own text for the show."

"Funny. Now back to Waterman."

"Well, he pretty much controlled this little domain before I came on board. I never suspected there might be something illegal going on, but I suppose there could be."

"So what changes have you made to disrupt his control?"

R. Z. Crompton

"I've made lots of changes in the production process, costs are down, quality is up. All those changes benefit him because the bonuses are higher."

"What if his benefits weren't from the bonuses?" By the time Rick had worded the question, they were at the back lot where the inventory of scrap had been piled for years. Rick's attention was diverted from his original question to a new topic. "Jesus, how long have you guys been building this warehouse of old metal? You expecting the melt down of America, and you'll be hired for the rebuilding?"

"I'm proud to say that our electric furnaces are the biggest recyclers in the state."

"I'd say you were the biggest stockpiler of junk," Rick harassed.

"You're right about one thing; this is a mess. I've been trying to get the scrap inventory under control ever since I started. We don't even know what's on the bottom of these piles; let alone the quality."

"Hum....Have you decreased the inventory since you've been here?"

"Yeah, why?"

"Maybe there is something in there he doesn't want you to find." Rick snapped another shot of the huge stock pile with his camera."

"I don't think it's that dramatic, maybe his scam is simply the scrap. The more I worked through the piles and cut things down, the easier it would be for the inventory to show a discrepancy with what is logged in the computer. Look at all the tire tracks." Kevin pointed at the ground around them. "There shouldn't be so much action out here. Everything is brought in and out by rail. Only the loader should be out here."

Rick snapped additional shots working his way around the huge mound of scrap. "Hey, c'mon around here."

"Yeah, what d'ya find?"

Steel Illusions

"Is this what you mean by a closed container?"

Kevin walked around the corner and looked at the cylinder laying in the ground. "Yeah." He answered bending down to inspect the object at his feet. Relief washed through him as he turned over the container revealing the huge gouge. "Only without the hole."

"Obviously. How about this over here." Rick pointed to something protruding from under a long sheet of scrap. He wasn't sure what the strange object was, but it certainly didn't look like it belonged in a junk pile.

"All this scrap is checked and weighed by Sam Mathers when it comes into the mill. I doubt if you'll find anything."

"I'd find out from Mathers when he started collecting wallets for melt down."

"A wallet?"

Rick bent down to pick the brown leather out of the heap. He opened it up looking at the license inside. "Do you have a Phillips working here, John Phillips?"

"I don't recognize the name. Is there anything else in there?"

"Oh, yeah. A couple hundred dollars and some credit cards. Wallet looks in pretty good condition, so it hasn't been here very long."

"I'll check in the front office to see if anyone has reported it missing."

"I doubt if it's been reported. Face it, if the guy wasn't supposed to be out here, why would he report his wallet missing. This just proves he was here."

"What could he have wanted out here?"

"You know the old saying 'One man's junk is another man's treasure'."

"That's bad. I can't believe you even said it."

"Think about it. If you don't know how much is really out here, why couldn't it be shipped in and trucked out for resale. Shit, you could pay for the same scrap two or three times."

"No way! There'd have to be too many people involved. People on the inside, people on the outside. I don't believe it."

"You've lived a sheltered life, my friend. Whoever set this little game up has had lots of time to put his network in place. When one part of the network breaks down, the broken part is replaced, or it may just disappear." Rick shook the wallet in his hand.

"You think Phillips is dead?"

"Murdered is my guess and, most likely, melted down."

"I don't believe it. How'd we get from an explosion to murder?"

"Via a kidnapping in Colorado. Did you call them?"

"Yeah," Kevin didn't even realize he was running his fingers though his hair until he felt the brittle ends. "There was still no word, but they suspect it was ordered from here. Someone who knows us."

"My point exactly. Who has the most to gain? The police use the money theory."

"I think we should call the police."

"Not just yet. One, you don't have any hard core evidence: a wallet and a theory aren't going to get their attention. Two, you may just end up scaring the bad guys off."

"What do you suggest? There's so damn many pieces, I don't know which trail to follow."

"The makings of a great story."

"It's a Goddamn novel; a fuckin' best seller."

Rick walked around the pile of scrap trying to come up with some kind of plan. "Let's try to put some order to our investigation."

"Our?"

"Yeah, you can't do this without me. You have trouble thinking outside your realm of ethics."

"You mean I can't think like a criminal."

"That's it. And I'm assuming I'll get exclusive rights to this story."

"If you can help get me out of this, you can have anything you want."

"Let's check our pieces: Pete is the first piece of this. He was on the floor without his coat. There had to be a reason, some kind of gain for him to take the risk. The explosion, the tracks out here, the wallet: all tied to scrap. Think about who has the control of the scrap. Who sees the stuff when it comes in? Who was in charge before you came? How would somebody truck it out? Have you tried to follow these tracks?"

"I don't need to. There's a back entrance to Green Country Recycling that goes in this direction. If trucks are pulling in and out of here, they're going through the back gate. The only other way in or out is the main entrance, and we have a guard stationed there."

"See what I mean about your thinking?"

"What?"

"You never considered the possibility the guard might be part of this scam, did you?"

"You're right." Kevin answered feeling slightly embarrassed. "Okay, how 'bout this. Mathers is the one who checks all the scrap when it arrives. He'd be a logical candidate."

"Prime. Would he intentionally let bad scrap into the yard?"

"I guess it's possible."

"We're making progress. Last time we talked, you thought it was impossible. So now we know the scrap could enter in via Mathers approval and then out the front or the back by truck. Now we have a possible reason for Phillip's presence."

"There's still no proof."

R. Z. Crompton

"Yeah, but we've got to know where to look to find the proof and that begins with understanding some type of motive and looking for the possible players. Now, what about the money? Who has the right attitude and enough power to control the players?"

"My first guess would be Waterman, but he seems too obvious."

"If you suspect him, you must have reason."

"Whatever happened to the theory that the most obvious suspect was usually innocent?"

"That's only true in the movies. Real life tends to be a little more complicated especially when money is involved. We can't afford to overlook the obvious. You need to hire a guy to replace Pete, right?"

"Yeah, eventually."

"Why don't you do a little name dropping. Tell everybody you're thinking about hiring Phillips for the job. See what the reaction is. Somebody around here knows the man; or, at least, knows what happened to him. Now, what about the source of scrap."

"I never thought about that. But you're right. Somebody on that side of the business has to be involved."

"So who are your players?"

"Green Country Recycling and United Metal Services are my main sources of scrap."

"So what do you think? Were they both selling to the mill when you came on?"

"No, only GCR. I brought UMS in about a year ago to give GCR some competition."

"That might give GCR the motive to get you fired."

"Yeah, except for the fact that Adam, who's with UMS, was out in Colorado. When I talked to him on the phone he was stuck in a small town east of Avon. And, come to think of it, that avalanche last night was just east of Vail. I don't know about his motive, but he certainly had the opportunity."

"Now you're getting the hang of this investigating stuff: means, motive, opportunity."

"What about Waterman?"

"Keep your eyes and ears open. If he's guilty, he may be dangerous. Remember somebody has already committed one murder, maybe more."

"I'm going to go back into the company files and compare the financial reports with the scrap inventory."

****

The two men headed back for the main office unaware they were being watched. Timing was just a little off. *Bradford should've left for Colorado already. What in the hell does he think he's going to find digging around in the scrap yard? The reporter's gotta go. He's stirring my pot, and it's in the wrong direction. I'll take care of him later. Now I'm going to have a sympathetic visit with the grieving Mrs. Hayes.*

The ringing of the telephone distracted the man at the window. "Hello."

"Did you see him and the reporter out there? He's sniffing where he shouldn't be."

"I know. I'll take care of both of them. Weren't you the one who said Bradford would be off to the wilderness as soon as his wife was snatched?"

"Maybe he doesn't know."

"He knows. He called it a little problem. When are you going back out there?"

"After I see Bradford tomorrow morning. He want's an accounting of the scrap that's been shipped over the last five years."

"Five years?"

"He's fishing. He won't find anything. I've made sure all the records match. What's your next step?"

"I'm going to check in on the pretty little Mrs. Hayes this afternoon. Call me before you leave tomorrow."

"What are you going to do with her?"

"First, I'm going to find out if she know's anything, then I'm going to, shall we say, comfort her."

"Don't hurt her."

"Give it a rest. As long as I'm in charge of this party, I'm going to enjoy myself and provide my own entertainment."

CHAPTER THIRTEEN

Laura woke up face down on the dusty wooden floor of a small cabin. Shivers ravaged her body, but she was afraid to move. Afraid of where she was and who she was with. Gradually the reality of the last hours came back to her. A rope had been tied around her waist, and then each end was tied around another person; she assumed men. Voices were muffled through the hats and face masks worn by her captors. She'd been given short verbal instructions to stop, go, and hurry. All night long she'd worked her way up and down the difficult slopes, slipping and sliding from tree to tree barely able to catch a full breath in the terrible wind. She remembered falling off a large boulder pulling the weight of the two abductors over the rock behind her.

"Damn you, woman. Watch where you're going."

You son-of-a-bitch, I hope you broke your neck. Laura thought the words, but concentrated on catching her breath while her captors cursed her lack of balance in the snow and wind. The darkness was never ending as hour after hour she focused on taking one step at a time. She was never asked a question or offered information. They had stopped intermittently for a quick drink of water, but she had been offered no food. Her toes and fingers had been numb for hours; but now, as she lie on the floor, they seemed to burn. She was sure she was suffering from frostbite.

They'd arrived at the cabin just after dawn. Blowing snow made everything seem white. Even though there was light, there was no direction. Standing was the only way she could tell the difference between up or down. The white was as blinding as the darkness. If she hadn't been pulled into the cabin, she'd have missed it completely.

Slowly, Laura tested her fingers and toes, flexing her tired frozen limbs from the tips inward toward her body. The burning

Steel Illusions

sensation continued to get stronger. Her feet had been tied together. Without moving her head she looked around one side of the cabin. There was very little to see. A table was against the wall, and she could see the rows of canned goods on the shelf. Whoever brought her here had clearly known the way and had planned on staying a while. She remembered thinking early on they'd be lost in the blizzard; but even before the snowmobiles were abandoned for the snowshoes, it was clear the leader knew exactly where he was going and how to get there no matter what the conditions. She'd been brought here for a reason, and none of the possibilities were good.

Light snoring caught her attention. Laura couldn't see who was sleeping unless she turned her head. Any motion made a scratching noise on the wooden floor. With great effort, she tried to control the speed of her movement. This might be the only chance she had to get a look at who had kidnapped her. Only one face was visible from her vantage point. She didn't know who the man was, but she did know it was the same man she'd seen at Cassidy's. The black dress pants and leather coat had been exchanged for a snowmobile suit, but she recognized the swarthy look he had about him. She turned away not wanting them to know she was awake.

Laura's mind began to wander as the minutes slowly passed. She'd asked several times about Krista. It was possible for the two of them to be mistaken. Her imagination played with her emotions as she lie quietly on the uneven wooden boards. What if they had taken Krista by mistake? What if they killed her or Krista in-spite-of her cooperation? Was she a fool for going along with them? Where's Kevin? Was he here, looking for her? Was anyone looking for her?

Think! Clues - little things. Adam said Kevin was looking for his gloves. Why? I never did find him. He didn't answer his phone or his pager. No! No way. He wouldn't do this to me. He loves me!

But, Adam was here. Maybe his meeting me was not by chance. Maybe I shouldn't have said no to him last night. Maybe his proposition was a test of some kind. No! I can't believe he's capable of this either. Her thoughts were interrupted by the movement behind her.

"Hey, Mac, get up."

"What? What's wrong?" The second man groaned as he became conscious.

"Blindfold the woman before she wakes up, then tie her to the chair in the corner."

"I'm surprised she's still alive. Most women can't make a walk like that."

"Yeah, but I was told she could handle it."

"I bet she can do a lot of other things too."

"Don't get any stupid ideas just yet. If Bradford doesn't play by the rules set for him, then you can have all the fun you want."

Laura tried not to flinch when Kevin's name was mentioned. At least now she knew her kidnapping had a reason and wasn't just a random event; however, she didn't know if she should be more frightened now or not.

"In that case, I hope he doesn't play. Hey, why do we gotta tie her up if we're gonna kill her?"

"I don't know, just do it. We gotta keep her alive for at least twenty-four hours."

"I don't have to listen to you. I found her; I'm the one who has the contact with the guy paying you, remember? You're simply the guide."

"And without me, you can freeze to death out here. Now blindfold her."

Laura assumed it was the man named 'Mac' who put the tape across her eyes and face. Then nearly jerking her arm out of her shoulder, he yanked her off the floor. She groaned as her sore frostbitten feet hit the floor. Mac threw her onto the small wooden chair she'd seen in the corner of the room. His short, jerky movements suggested to her he was about to lose his temper, and she hoped his anger would be directed at his partner and not her.

"When you're done with her, go out and bring in some firewood."

"Do it yourself." Mac snapped as he pulled the rope tight around Laura's wrists and tied her to the chair. "Earn the money you're being paid."

"I have something else that needs tending to. The wood's in a shed around the corner to the left. We can have a fire as long as the weather is bad. As soon as it clears, the smoke might be seen."

"What do we do then?"

"Don't worry. I've been hiding out up here for years."

Mac accepted the fact he couldn't survive without Jake, so he reluctantly went outside to fetch the needed firewood.

As soon as the door closed, Jake was standing in front of Laura. She smelled the stench of his tobacco breath when he moved in close to her ear. "I have something special for you Mrs. Bradford."

There was no way for Laura to pull away from his touch. Mac had made the verbal threat against her body, but this man was getting up close and personal. Too close. He removed the scarf she'd had tied around her neck and unfastened the collar of her snowsuit all the time breathing the heavy fragrance of chewing tobacco in her face. The slow, deliberate movements were loaded with pleasure. The large, silver zipper of the cumbersome snowsuit was pulled down past her waist allowing the suit to be slid over her left shoulder. For just a second there was a pause in his abusive touch.

"This will help you relax a little and get the rest you need."

R. Z. Crompton

Laura wasn't sure what he was talking about until she felt the sharp sting of the needle in her upper arm. Within seconds she was groggy, and she felt her head slump to her chest.

****

Megan gasped for a breath as she looked around her. She didn't realize until she stopped moving she'd been holding her breath. Now she understood what Luke meant about fear. Megan had been so determined to keep her mouth shut, she'd forgotten to inhale through her nose.

That wasn't so bad. Kinda like a mega carnival ride. Megan fumbled with the clasp of the helmet which had protected her from injury. She pulled it from her head and threw it to the ground. "Hey!" She screamed. But there was no answer. *Shit where is everybody?* "Hey! This isn't funny! Come dig me out."

A lone glove stuck out of the snow a few feet from where Megan was buried. It was then she realized everyone else was entombed under the sea of white. *Holy mother of God! They're all gone.* "Krista! Brad! Luke!" She heard herself scream the names, but there was no answer.

Don't panic. DIG! Just Dig. Megan gave herself the same advice her mother would have, then set to work. She forced her hands to dig as fast as possible, but the snow was hard packed and difficult to maneuver. The pole, she'd used as a walking stick, had been lost in the slide; so there was nothing to help her move the heavy chunks of snow. The sweat ran down the side of her face and soaked the clothing next to her skin, partially from fear and partially from the strain. She'd been buried horizontally in about eight inches of white, icy concrete. Digging around her waist, she tried to get some leverage in order to pull her legs out. Finally, Megan was able to twist her hips around so she could use her arms to heave her way out of the gorilla like grip.

Megan unzipped her pocket taking out the beacon. The tiny "send" light was still on. She switched it to "receive" and high pitched beeps came from every direction. *Shit! He told us to turn them on, but not how to use them. This isn't going to do me or them any good.* She limped over to the spot where the tip of a glove was sticking out of the snow and started digging. The hand in the glove acknowledged her presence by gripping her fingers when she reached for it. The sensation of life buried beneath the snow struck a cord of horror in her soul. She had no idea of who was down there, but it was a friend or sister grabbing for life. Megan almost had to pry the fingers lose, so she could start digging.

Intent on getting this body freed from the snow, Megan didn't hear Stewart thrashing through the trees. "Meg!" He nearly scared the life out of her. When she looked up, she saw a weary and frightened friend, dragging his helmet and backpack, heading toward her.

"Jesus, Stewart. You shouldn't sneak up on a person that way. What happened? Why weren't you caught in the slide?"

"I felt the slight movement of the fracture while I was still close enough to grab a tree. I held on while the rest of you were swept away. There was nothing I could do until the snow passed me, then I high-tailed it through the trees looking for you." He flung his pack to the ground and rummaged through the content.

"I'm really glad to see you."

"You hurt?"

"My leg aches a little, but I think it's just a bruise. I can't work this beacon."

"You dig. Here's the shovel from my pack. I'll find the next body...." Megan shot him a glance, and dug faster. She didn't want bodies. She wanted her friends and her sister.

R. Z. Crompton

Luke had been the closest to her, and it was Luke whose grateful face appeared to her about eight inches under the snow. Within minutes, she had freed his upper body enough so he could help her dig. "I've got a shovel in my pack, too. Get it out for me."

"I found somebody. Hurry!"

"Go, Meg. Try to free the snow around the face first. I'll finish digging myself out. As soon as Stewart locates another body and you've dug air space, move on."

"Yeah," was all Megan said as she hurried to the spot Stewart had marked as another tomb.

Luke was out in minutes and turning his beacon to "receive." Working together, he and Stewart were able to find the next spot in less than a minute. They worked in a grid pattern closing in on the signal coming from one of the transmitters.

Megan didn't worry about who was under the chunks of snow; it didn't matter. She was desperate for time, and it took several minutes to uncover Brad's head and give him some freedom of movement. She removed his helmet for him which made breathing easier. "Thanks, Meg. I'm sure glad to see you. Krista, where's Krista?" He asked gasping for air.

"I don't know. We haven't found her. You feel okay?"

"Oh yeah, ready to dance." His attempt at humor didn't hide his fear. Megan could see it in his eyes.

"Luke and Stewart just found somebody else, Brad. I gotta go help. We'll be back."

"Wait. Help me get out."

"I can't, I have to get air to the other body they found. Okay?"

"Go. I'll count sheep 'til you get back."

"Good boy." Megan forced herself to move to the next spot. The exertion from the fall and the digging was taking its toll on her. Her left leg throbbed with pain as she braced herself to start digging

Steel Illusions

again. Luke and Brad had been buried under only eight to twelve inches of snow in a horizontal position. Megan had been surprised the guys couldn't dig themselves out, but the snow was so incredibly heavy. If they had been any deeper she didn't know if she'd have had the energy to get them out. Luke had left a glove to mark the spot where Megan should dig and then moved on. By this point her breathing was labored as she tried to carve a hole out of the newly formed glacier. It was the sight of long brown hair which caught her attention and gave her new energy. The hair belonged to her sister. "Krista," Megan whispered. "Krista!" Yelling her sister's name didn't help. There was no movement as Megan cleared the area around Krista's face. She'd lost her helmet in the slide.

Megan dug around the head and upper body at a frantic rate. The pain in her leg had disappeared. Her sister could have been fatally hurt in the fall, and there'd be nothing she could do. Megan freed the chest area and maneuvered herself into a position to give mouth to mouth. She tried to lift Krista's head back, clearing the airway. The body showed no movement. She pinched the nose, took a deep breath and blew the life giving oxygen into her sister. Nothing. She blew again.

Luke knelt beside her quickly trying to free more of the snow from around Krista's body. "Let me help." He cleared the area around Krista's chest so they could both get closer to her, but her head was below her feet not only in the direction it faced down hill, but also in snow level. The positioning of the body was awkward, but they couldn't waste time digging her completely out of the snow.

Megan was out of breath; she moved out of the way so Luke would have the best angle. "Luke, there's no pulse. We gotta do CPR. You blow; I'll compress." *Dear God, save her. Please save her.* "One Mississippi, Two Mississippi, Three Mississippi............"

Stewart was left to dig Chelli out of the snow. Her helmet had also been lost in the ride down the mountain. She was breathing, but barely. She had a large cut and bruise on her forehead. She had come to a stop face down; so even though there was only a few inches of snow covering her, breathing would have been nearly impossible. Stewart worked skillfully digging around her body. There was a deep groan from her throat as he finally freed her from the snowy bondage and gave her a drink of water.

Poor Brad had been left in the snow the longest. Patience had never been his strong suit, but today he forced himself to stay quiet. They'd be back, and he needed them to find Krista. At least he knew he was alive and relatively unharmed. The hardest part about waiting was not knowing what was going on. It seemed like forever before he saw Stewart's face hovering above him.

"Boy, I'm glad to see you."

"Yeah, I know. I'll have you out as quickly as I can."

"What about the girls? How are the girls."

"Not so good. Let me free your arms, then you can help dig yourself out. We need you to help dig a shelter. The girls aren't going to be able to go anywhere for a while."

"Are they both alive."

Stewart didn't want to answer. He wasn't sure. Luke and Megan were still trying to revive Krista. "Let me get you out of here."

Brad was able to move his head and upper body within minutes. He twisted around and saw Luke and Megan kneeling over a body. "Who are they working on, Stew?"

"It's Krista."

Brad systematically chopped away at the snow trying to free his waist. The flow of adrenaline and sheer desire allowed him to pull himself out of the snow. He crawled to the inert body laying on the snow.

Steel Illusions

"Is she....?" He couldn't ask the question.

"No." Megan answered with a sigh of relief. "We finally got a weak pulse, and she's breathing on her own. She's not conscious. How are you?"

"Sore, but moving. How 'bout Chelli?" Brad watched as Stewart tried to make her more comfortable.

"Don't know how badly she's hurt. At least she's aware of what's happened to her. I think Chelli was buried too long and her body temperature dropped. We've got to get them into some kind of shelter and warmed up, or they'll both go into shock." Luke looked around for a good place to set up a camp.

"How far to the cabin?" Brad asked hoping they could still make it.

"It's not that far. The distance isn't the problem; it's the terrain. Can you walk?"

Brad tried to stand as Luke looked around the area for a safe place to build a shelter. "I'm not going to be able to get very far. Must'a twisted my ankle. I can dig if you show me where."

"Let's try over there." Luke gave Brad an arm for support as they stumbled through the uneven hunks of snow.

Megan spoke softly to her sister, "C'mon, you have to wake up. Krista, I know you're in there. This is no time to play hide and seek. I'm going to tell Brad all your secrets if you don't open your eyes." She chided trying to get a few drops of water into her sister's mouth.

Slowly the eyelids fluttered open, and Krista smiled as she looked up into her sister's eyes. "You better not tell. I'll beat you up."

"That's better. I like my sister with a little spunk. Now drink this. Just a little at a time."

Krista tried to drink and talk at the same time, but Megan anticipated the question. "Brad hurt his ankle. He's helping Luke and Stewart dig us another snow cave."

"That sounds pretty good right now. I'm so cold. Chelli? How's Chelli?"

"I think she's suffering from hypothermia. I'm going to check on her, you stay calm. Don't try to move around. I don't know if you've got internal injuries."

"Okay, Doc." Krista could see nothing but the grayish sky. Snow swirled above her face until a gust of wind ushered the flakes away. There was no particular pain. She just felt numb. As a second year med student, she knew enough to take Megan's advice seriously. Krista was terrified she might have injured her neck and back.

"Chel, you still with us?" Megan questioned softly as she approached the still body.

"Yeah, I'm freezing. How much longer?" She could barely speak through her chattering teeth. Megan could see the violent tremors of cold vibrate through Chelli's body.

"It won't be much longer. Let me see if I can find a sleeping bag for you. I lost my pack in the snow, but Luke still has his. Hang in there, Chel." Megan looked for Luke. She was surprised at how quickly the three of them had dug out a shelter. "Luke, I need to cover Chelli and Krista with sleeping bags until we can get them inside."

"Be careful. Hypothermia victims have to be warmed from the chest area first. Warming the legs and arms too fast can cause cold blood to flow into the heart and lungs."

"Luke, how are we going to move Krista?"

"Is she conscious yet?"

"Yeah, but she said she can't feel anything."

"Give her a few minutes. It's not so unusual to feel numb after a shock to the body. We're almost done here. Take my pack and use what you can. We'll look for the missing ones after we get the girls settled in the cave. I think Stew has the floor mat in his pack. As soon as we get some air into it we'll move the girls off the snow." Megan walked back to nurse Chelli and Krista.

"Hey, Stew, did ya hear that?" Luke questioned.

"Already found the mat. Brad, take a break. It'll be easier on your ankle if you blow up the mat while Luke and I dig."

Brad was glad to sit idle for a moment. The pain in his leg was becoming more acute with every motion. The girls were not the only ones who couldn't finish the trip. He didn't think he'd have the ability to get himself into the cave. "Here's the mat. How much longer before the cave is dug out?"

"It's not big enough yet." Luke responded. "Let's move Chelli and Krista onto the mat. They'll feel much better when we get them off the snow. You're lookin' pretty pale, Brad. You better sit still." Luke and Stewart headed for Megan and her two incapacitated patients.

Luke laid a sleeping bag next to Chelli. It was the only thing handy to use as a gurney. Carefully, Chelli was placed onto the bag and carried to the mat. They returned for Krista.

"How ya doin'?" Luke asked kneeling beside her. If she'd fractured her neck or spinal cord, they could do more damage by moving her.

"Actually, I feel better. Every bone in my body aches like hell, but I think I can move if you'd tell my hovering sister to get out of my face."

"Oh, she's got an attitude, too. That's good. But I agree with your sister, you might as well let us give you a lift as long as we're here." Luke and Stewart placed their second victim on the makeshift transport and moved her to the mat. Both girls shared a sleeping bag and one was given to Brad, who was helped onto the mat beside Krista. Luke was afraid he was going into shock also. In fact, he looked worse than the girls.

"Megan, start up our little fireplace. The can should be in Stew's pack. Boil a pan of water, so we can make some extra thick hot chocolate. The extra sugar will help replace the energy we've lost from the shivering and over exertion. I'm going to see if I can locate the two missing bags. We're going to need them. The first aid kit was in Chelli's pack. Stewart, you want to dig or look for bags?"

"You look; I'll dig."

Megan was left to nurse the injured and didn't mind. "Brad, how's that leg?"

"I think I twisted my ankle."

"Mind if I take a look at it?" Megan helped Brad loosen the ties on his boots and work the pant leg up so she could see the source of his pain. "My God, Brad, what in the hell were you doing walking around on this leg? It's broken. In two places, I think. No wonder you're in so much pain."

"It must've been the cold. I really didn't feel so badly when I started digging the shelter. I was more worried about Krista and Chelli than myself."

"Krista, I don't know what to do for this."

"What do you do with one of your animals?"

"Unfortunately, we usually put it down."

Chelli wasn't exactly sure what her words meant. "How do you put a person down?"

"Chelli, that means we kill it."

"You can't kill him." Chelli gasped

"No shit!" Megan and Krista laughed at her naivé remark. Even Brad didn't pass up the opportunity to snicker. "Chel, did part of your electrical system short out in the snow? Krista, any suggestions about what to do?"

Steel Illusions

"Is he bleeding?" Megan shook her head negatively. "Good, keep the leg slightly elevated and him warm." She paused before adding, "and ask Luke as soon as he comes back what you should do."

Megan finished purifying the water and adding the instant chocolate mix. It cooled quickly to a drinkable temperature.

Stewart announced completion of the domed shelter about the same time Luke came back with one of the missing bags. Chelli, Krista, and Brad were tenderly moved around until everyone was neatly tucked into the cave. Krista was feeling better and took some aspirin for her minor bumps and bruises. Krista was pretty sure Chelli had a concussion. Her pupils were dilated and she complained of an upset stomach. Brad was nearly unconscious. Both of them needed medical attention immediately.

Luke took the small black radio out of his pocket. He patted himself on the back for remembering to zip the thing into his pocket just before they'd crossed the first clearing. He'd have a hell of a time contacting Tim if he'd lost it in the slide.

"Team #1 to base. Anybody home?"

"Hey, Team #1. How ya doin' up there?"

"Not so good. We got caught in a slide. Two seriously injured. How long before you can get a chopper up here. We're at the edge of the clearing about two miles northwest of the cabin."

"Sorry, no go. The winds aren't expected to die down enough to make a landing 'til morning. Can you last?"

"Do we have any choice?"

"Not really. Stay warm friend."

"Any news about the Pass or Mr. Bradford?"

"Bradford called in. He's fine, but a big explosion at the mill makes it impossible for him to come out here. A man was killed. We told him there was nothing he could do here but wait, and we'd call as soon as we had any news. He's got reservations on several flights, so he can be here as soon as we find Mrs. Bradford. Any sight of them?"

"We found a possible trail just before we were caught in the avalanche...."

"Luke?" Megan interrupted, "Luke?" She grabbed onto his arm.

"Hang on, Base. What Meg?"

"Ask if they can air drop in some supplies and get a doctor to the radio to help us."

The question was asked, and a doctor would be available in about twenty minutes. The supplies was another matter.

"The chopper took a couple of seriously injured folks from the slide into Denver. We don't have anybody else. Besides, Luke, even if the chopper was here, I don't know if anybody would be able to find you in this weather."

"I know somebody. You find Jones; he knows every inch of this place as well as I do. I also know he can fly that plane of his through a hurricane if he has too."

"Steve Jones? Are you sure he won't bite my head off if I call him."

"No. He's a crusty old goat, but inside he's really soft as a loaf of fresh bread. And he was my Dad's best friend."

"If you say so. Can you put out any kind of marker?"

"Can't send up a flare; we're too close to the cabin. I'll set out the orange flags in a circle. That might help."

"Okay. I'll take care of it. By the way, you asked about the pass. Cars in the west bound lanes were buried under twenty-five feet of snow. Ten dead. Most of the people were shaken up, but not injured. It was set intentionally. Base out."

There was silence as everyone digested the tragedy of events.

Steel Illusions

All the clothes were damp from the sweat and snow. Chelli was undressed and tenderly moved into a sleeping bag. Stewart found dry clothes in one of the packs; he changed and crawled in beside her. He used his body heat to warm her. Krista took another set of dry clothes and cuddled up next to Brad. He had to be kept warm. Both he and Chelli drifted in and out of sleep.

Megan grumbled under her breath as she shoved some hot chocolate under Krista's nose, nearly spilling the thick liquid in her lap.

"What's wrong with you?"

"I can't believe Dad isn't coming out here," Megan snarled at her sister.

"There was an explosion; a man is dead."

"Mother's more important than the mill. He always put that damn mill first."

"Meg, you're exaggerating just like Daddy does."

"No, I'm not. The mill was down - he missed a concert. The mill was down - he missed a soccer game. The mill was down - he missed a vacation."

Krista didn't interrupt. Part of what her sister was saying was true, and it was better to let Megan vent her frustrations, so she didn't take her anger out on the rest of them.

"His Globetrotter Meetings were more important than vacationing with his family."

"Dad loves us; you know that."

"When it's convenient."

"Stop it! He provided us with a good living. He did his job and Mom did hers. She was there chauffeuring us everywhere, pushing us. We wouldn't be at A & M if she hadn't been there encouraging us - helping us."

"That's my point. She was there, and we're here now looking for her. Where's Dad? With his Goddamn mill."

"I beg your pardon. Mom wouldn't like your language! By the way, that mill has paid for your education: all of it."

"I don't think so. I've paid for my education with my scholarships."

"You ungrateful wretch. How do you think you got all those scholarships. It was the voice lessons, clarinet lessons, soccer, the books, the camps, the trips to France. Mom made sure we did it, but Dad paid for it. They were partners; and without both of them, we wouldn't be what we are."

"All I know is that Mom's in trouble, he's not here." Megan was nearly screaming at her sister which caused Krista to respond in kind. The others sat quietly by and watched in amazement as the two girls aired their emotions.

"What would he do? Sit in the rescue office and twiddle his thumbs?"

"Yes! That's where he should be."

Luke listened intently to the arguing before interrupting. "Meg, what if he can't come?"

She nearly bit his head off for intruding on the family fight. "What are you talking about?"

"Well, there was a fatal explosion the day before your mom was kidnapped. Do you really think it's just a coincidence?" Luke asked her softly, hoping she wouldn't start yelling at him.

"I never thought of that. But what difference would it make? The explosion is over. The guy is dead. And we need him here."

"Luke's right, Meg." Krista said more calmly. "Somebody was betting on Dad's leaving as soon as he heard about Mom. That'd be his natural reaction. Don't start shaking your head. You know I'm right. I bet it's killing him to not be here."

"If he leaves, he'll be playing the game by rules he doesn't understand. If he stays there, at least he has a chance of finding out who did this." Luke offered his most reasonable explanation.

"My God, I never thought about it; but ya know, just because we find Mom doesn't mean we'll find out who planned this." Megan seem truly surprised at the possibility. "I'm sorry I blew up."

"Don't worry. I'm scared too." Krista confessed. "What now, Luke? You're our fearless leader."

Luke shook his head. He wanted to get to the cabin, but leaving the injured behind didn't set well with him. "I don't know."

"We have to go on, Luke." Megan pleaded. "If we don't, we lose the surprise."

"She's right. We're almost there." Stewart added from the sleeping bag.

Krista wasn't silent either. "I'll take care of Chelli and Brad. If you wait 'til morning, the weather will clear. They'll be watching for you."

"It's not good to split up a party. You know that, Stewart." Luke was questioning his own wish to continue the trip.

"Maybe it's okay if you're trying to save a life. We can go down to the cabin, rescue Laura and be back here for dinner."

"Stewart, it's not that easy. What are you going to do with the bad guys when you get there?"

Luke had reservations about trying to overpower somebody with half of his team held up in a snow cave, and he was worried about the welfare of those left behind.

Krista was afraid he wouldn't go and all of their effort would have been wasted. "I can talk to the doctor as well as either of you can. I can nurse both of them and take turns keeping them warm. The blankets and dry clothes will help. As soon as the supplies get here, you go."

"No." Luke was firm in his answer. "We don't know how badly you're hurt. I don't want you moving around trying to nurse them when you need nursing yourself."

"I'll stay with them," Megan offered. "You and Stewart are the strongest."

"I don't like that idea either," Luke protested. "Stewart, you've had the experience to deal with this kind of weather and the first aid training. We don't know how long it'll take Steve to get up here. If you're willing to stay, Megan and I could leave right away. We'll take one backpack with just the bare necessities and try to get back before dark. Is that okay with you, Meg?"

"Works for me," was all she needed to say.

"How 'bout you, Stew?"

"Sounds like a plan."

Megan reached over to give Krista a hug. "You take care."

"You too. Bring Mom back with you."

CHAPTER FOURTEEN

Megan and Luke plowed through the chunks of snow in the clearing where the avalanche had tried to destroy them. Mother Nature had been kind today yielding up the victims of her fury with merely a warning of her power. Leaving the injured huddled in the snow cave was the remnants of the near disaster. When Luke and Megan took their last look at the friends and members of the rescue team who were being left behind, Brad was in the most serious trouble. Using the thermometer in the medicine kit, Krista had realized his temperature was alarmingly low. Her own body heat was the only thing she had to keep him warm

Megan, concerned about her ability to be of much help to Luke when they arrived at the cabin, trudged quietly behind him. If nothing else, she could offer him company and morale support on the trip down the mountain. After a couple hundred yards, Megan picked her topic and spoke into the microphone.

"Luke, didn't the police ever have any suspects?"

"What?" He was surprised by her question and wasn't sure what she was talking about. He'd been deep in a mental conversation with himself about what to do when they reached the cabin. How could he possibly overpower two men? There had to be at least two maybe three since there were two snowmobiles missing. Unless Laura had gone away willingly, he was sure she'd been tied to the machine behind one of the kidnappers.

"Your grandfather's death. Were there no suspects?"

"Oh. Not really. We weren't able to come up with anyone who had a grudge or a reason to kill him, so everybody just assumed it was a poacher who'd caught Gramps by surprise."

"Do you miss coming out here? I know it's beautiful in the summertime."

"I didn't know you'd ever been up here."

Steel Illusions

"Yeah, we used to do some horseback riding when we came out in the summer. In fact, I think we used Steve Jone's Stables at Beaver Creek. Isn't he the guy who's going to fly in the supplies?"

"I hope. My positive tone was for the benefit of the guy on the other end of the radio."

"Do you think he might say 'no'?"

"I doubt it, but I'm sure he'll grumble a lot. It's in his nature. I enjoy his crotchety old attitude especially since I know he's got a heart as big as the mountain he lives on." Luke changed the topic. He'd never been comfortable talking about himself. "So how come you always came out here in the summer? Most teenagers prefer going to the beach." He puffed as he finished his thought. Moving through the deep snow on steep terrain was hard enough without trying to carry on a conversation. He could hear Megan's struggle to breath through the microphone.

"Oh, we did from time to time. But the mountains always called us back. There's more to do out here. I can only lie on the beach so long before I need something to do. We always enjoyed hiking. In fact, we started hiking the 14'ers when I was fifteen. My Uncle Dennis took us out to conquer Grays Peak, and Dad nearly died."

"Really?"

"No, not really. But the altitude was more of a challenge than the mountain. Fourteen thousand feet is just a little thin on oxygen."

"I mean about his being with you. I thought he never left the mill."

"Okay, maybe I exaggerated just a little. I remember feeling totally overwhelmed as I looked at the spot where a ranger pointed out a recent winter avalanche. The trees must have been snapped like match sticks under the force of the snow. They were scattered along the obvious path all the way to the creek bed at the bottom. Dad tried to hurry us along by telling us it'd be warmer at the peak because we'd be closer to the sun."

"Did you believe him?"

"I wanted to; it was cold up there. We had a snowball fight on our way down. We even saw people with skis strapped to their backs hiking the four miles to the top, so they could enjoy the excitement of jumping off the peak and skiing half way down. I was amazed. Have you ever done something so crazy?"

"Would you think me a total idiot if I said yes?"

"Really! Have you done that?" Her excitement caused an instant lapse in concentration, and she stepped out catching the back of Luke' snowshoe.

Luke's forward motion was instantly hindered by being anchored to the ground. He nearly went head over heels down the slope. "Ahhh..." was the reaction as he reached for the snow covered trees next to him. Megan's body was thrust backwards when Luke's momentum pulled his snowshoe out from under her.

"Ouch!" She yelled, landing backside first on a large boulder. Feeling very silly and clumsy, she stood up brushing the snow off from the waist down. "Sorry, Luke."

Luke's arms were wrapped tightly in the branches when he finally stopped his forward lunge. Thanks to the face shield on the helmet, his face was protected; or he'd be enjoying a healthy dose of greens.

Megan thought she heard low grumbling coming through her headset. She deserved to be cursed, but just as she was about to beg forgiveness for her lack of gracefulness, she realized Luke was laughing. "I'm sorry." She said more distinctly. "Are you okay?"

"Don't worry." There was a pronounced sense of humor to his voice. "That hasn't happened to me for years. I did the same thing to Gramps when I was a kid. Sent him face first into a bunch of trees.

Steel Illusions

We didn't use helmets back then, so he came up spitting pine needles like a wild cat. Ya know, now that I think about it, I know why I enjoy Jones so much."

"He reminds you of your grandfather?"

"Yeah, he does."

"Speaking of wild cats, I was going to ask you if I needed to worry about wild animals."

"Not really. The bears are all asleep, and they don't want to see you any more than you want to come face to face with one of them. We rarely see big cats around here. Most of the wild animals are relatively harmless."

"What do you mean by relative?" Megan adjusted her snowshoes before following Luke down the steep trail.

There was a pleasant lightness to his tone when he answered. "This time try to stay a little further behind me."

Megan was glad he couldn't see her flushed cheeks.

"Relative is in the eyes of the beholder. Some of the animals, raccoons for instance, can destroy your camp if you don't keep your food and supplies picked up. If your life depends on those supplies, you could be in big trouble. A porcupine can strip all the bark off a tree, and the tree dies. Actually, the most dangerous animal in the forest is us."

"I know. We destroy what we don't appreciate, and most people miss the simple beauty of nature."

Luke was impressed with her respect for the mountains. "That's why we have designated wilderness areas. No motors are allowed in those areas. It cuts down on the traffic. Not as many people are inclined to traipse through an area on foot or horseback as they are in a four wheel drive."

"What about snowmobiles? Don't people cheat sometimes and bring the snowmobiles back here?"

"Yeah, but most of this terrain is too dangerous even for the snowmobiles. I prefer the wide open bowls for riding, how 'bout you?"

"Actually, I prefer a Harley on the highway."

"What? You ride a motorcycle too?"

"Not exactly. My aunt and uncle both have Harleys. Only my uncle can take a rider, so Krista and I always argue about whose turn it is. The ride is fantastic. All I have to do is keep my mouth shut and enjoy the vibrations."

"What?" Luke tried to hold back his laughter.

"You know. The bugs. If you don't keep your mouth shut, you could end up with a high protein lunch."

"I was questioning the 'vibration' part."

"Oh. Well, what can I say? It's better on a Harley."

"I've never had the experience."

"Probably not," was all she said, but there was a smile on her face as the picture of riding through the Black Hills on the back of her uncle's Harley played across her mind.

Once again they had found a slow progressive pace reaching from tree to tree. "Luke, you never answered my question about skiing off the top of a mountain."

"I think it probably ranks right up there with whitewater rafting: totally nuts. But, yes. I'm guilty."

"Oh, whitewater rafting is absolutely my favorite! I love the rush of adrenaline as we fly through the class three and fours rapids. I could do Dead Cow on the Lower Eagle and Sink Hole on the Arkansas every summer and never get enough."

"You sound like a commercial."

"Don't you enjoy the rivers up here? Just getting in the water is a rush. It's so cold. Like ten thousand needles jabbing at your body."

"I enjoy the fishing more than the rafting, but it has it's moments, I guess. Who's your outfitter when you're here?"
"We usually go with Lakota tours."
"How come?"
"We like to stay at Montaneros at Lionshead. Lakota tours is just outside the door. It's easy to make the arrangements, and they offer a variety of choices."
"Montaneros is nice. Tim works there."
"I knew I'd seen him somewhere before. I just didn't recognize him in his different role. Tim's the guy who suggested Lakota tours to us a few years ago.
"Good suggestion. I know their guides are experienced. I'd probably use them, too."
"I don't know about experience, but they are certainly handsome. Tom was our guide last time we went, and he was great on the river and..."
"And what?"
"Let's just say his bed side manner was pleasant."
Luke was silenced by her comment. He'd hoped his assessment of her had been wrong. His lack of comment left her time to continue.
"Even Dad enjoyed Tom's sense of humor."
"Your Dad?"
"Yeah, Dad was with us again. I guess I really did exaggerate about his absence."
"You didn't go out with Tom, or whatever his name was?"
"Of course not. I was fifteen, and he was on his way to medical school in Wisconsin."
"Sorry, Meg. I owe you an apology. My imagination and assumptions keep getting in the way of my usually, pretty good judgement."

"Don't worry about it. I suffer from the same malady." Megan brushed the words off graciously, and continued her testimonial about the glories of whitewater. "What are your favorite rapids?"

"I don't know." Luke answered, sounding surprised at his ignorance about such a popular sport.

"Are you telling me you've never been whitewater rafting?"

"My mother wouldn't let me go when I was young; and when I was old enough, I no longer had the curiosity."

"That's like living near one of the Wonders of the World and never going to see it."

"You're exaggerating again." Luke was smiling, but she couldn't see his face behind the face shield. Her excitement was contagious, and he appreciated her love for his backyard, and the fact that he was able to distract him from the rigors of the hike.

"This summer, we're going to have to fix the problem. How can you know so much about the desires of the people you serve and not appreciate the raging water?"

"I guess I never thought about it. When I was younger, there wasn't time; and now that I'm older, there still isn't time."

"That's ridiculous. Are you afraid of the water?"

She certainly didn't beat around the bush, and he wasn't sure of the answer. "I don't know. My mother was terrified of everything up here. She never learned to love the wilderness like my father and I. Maybe she passed some of her fear to me. I know my sisters certainly share her feelings. They want out of this valley."

"That's too bad. Have they never been to Beaver Lake and looked through water so clear it seems the fish are only below the surface, or watched the deer graze in the meadow? I remember the baby chipmunks scurrying to take a chip tentatively out of my hand while I sat quietly by the water."

Steel Illusions

"I don't think either of them have ever been on horse back, and they certainly wouldn't hike. Mom made it clear to them that the only way out of here was to get a good education."

"What are you going to do when they're both gone? Are you still planning to leave?"

"Not anymore. After I came back and got involved in the business, I realized I'd never leave for good. In spite of the sadness I experienced here, it really is the only place I'm happy."

"I know what you mean," Megan said breathlessly as she reached for a tree limb to brace herself on. The huge boulders were difficult to move around and staying in an upright position took nearly as much energy as moving down the side of the slope.

She took a deep breath and tried to continue. "The mountains are wonderful in the winter, but they're glorious in the summer. The various shades of green mingled with the bright colors of the wild flowers against the deep blue of the sky soothes my soul every time I'm here. I like the wind whispering to me through the aspen trees, and the delicate purple and white Columbine floating on their long green stems."

"If you enjoy the area so much, why don't you move out here?"

"I have to finish school first. My dream has always been to have a couple horses, start my own vet clinic and live in the mountains."

"Sounds like a good dream to me." Luke hadn't noticed Megan falling further and further behind. Her voice still coming through the microphone loud and clear made him think she was right behind him. In fact, he'd found himself looking from time to time to make sure she wasn't going to step on him again. He was surprised to turn around and not see her. "Megan?"

"Yeah?"

Just then she came into view with sluggish movements giving her a pace much slower than his. "Why didn't you tell me to slow down?" He stopped now to wait for her.

"I kept hoping I'd catch up," she puffed, bracing herself against the large snow covered granite. After a few deep breaths, she asked, "What are we going to do when we get to the cabin? Any ideas?"

"No. Not yet. Any suggestions? You seem to work well under pressure."

"What do you mean?"

"You were the one so terrified of being caught in an avalanche, and yet you remained calm during the whole incident. I was impressed."

"I guess when it came right down to the moment, I didn't have time to think about the fear. There was only time to react."

"Yeah, well some people react to their fear not the situation. One year I watched a man run screaming down the interstate after we dug his car out from a slide. The irony was that only the back end of the car was buried. He could've gotten out the front door, but his fear had him paralyzed. You'd make a good rescue worker if you ever move out here, especially with your medical background."

Megan laughed at his remark. "Medical training? I'm going to be a veterinarian."

"Out here in the wilderness, medicine is medicine. Actually, we've been talking about using dogs in our search and rescue operations, but we don't have anybody qualified to work with the animals. Maybe that's your calling."

"Who knows?" Megan shrugged off the idea. Even though it sounded good to her, this wasn't the time to make plans for the future. "Back to my question. What are we going to do when we get down there?"

"What any good rescue team does. Give me your best guess. What would you do?" Luke asked wondering if she'd really come back to the mountains.

Megan was afraid he'd say call in for back up, but she had a gut feeling it was the wrong answer. A good rescue team shouldn't have to wait. They'd be trained to react to the situation or a person might die.

"I would wait until I got there to evaluate the situation. We may have several options depending on how many people are involved and the conditions around us."

"See, I knew you'd be good at this." Luke was being genuine in his compliment. Not everyone could act as calmly and intelligently under stress as this woman did. "You can be my partner anytime."

"Thanks, but maybe you'd better wait until this is all over before making a judgement like that. I'm not in as good of shape as I thought. This hike is about all I can handle." Megan hadn't realized how much terrain they'd covered during the light conversation.

"It'll be over before you know it. The cabin is just off to your right a couple hundred feet. Stay close to the trees. If somebody does look out the window, we don't want to be seen."

****

Laura felt a strange warmth surrounding her body. She cuddled closer to the pleasant sensation. Softness stroked her body. She'd been so cold. The last thing she remembered was being terribly cold, but now it was cozy, safe. When she tried to roll over, reality broke through the sleepy haze. Her arms and legs were tied to the corners of the bed. Only a slight movement was needed to confirm her fears. She'd been stripped of her clothing and placed between several animal furs. An unusual warmth radiated from underneath the soft covering.

R. Z. Crompton

It must have been the source of her comfort, which she couldn't deny was better than being tied to the chair and freezing to death. However, she'd have been a fool to think that these guys were concerned about her comfort.

Laura's memory was still foggy. She remembered the men arguing, and Jake. He was the scary one. The other guy, Mac. He was all talk. Laura had no doubt it was Jake who'd undressed her. He seemed very calculating and sure of himself out here. This was his playground, and he could do anything he wanted.

As the drug wore off and her thinking became clearer, images filtered through her mind. She'd been dreaming. Dreaming about Kevin? She wasn't sure. *It must've been Kevin. He was stroking me. Kissing me, loving me. No, it wasn't Kevin. Was it Adam?* Reality was slow, but not slow enough. She'd been better off with no memory of what had happened. Tears filled the corner of her eyes as she realized she'd not been dreaming. Jake had raped her. His face filled the space where Kevin's should've been. *Hell, even Adam would've been a better option than this. What now?*

"I see you're awake. It's about time."

Laura turned her head to look at the man coming through the door with an arm load of firewood. She recognized the face. It was Jake's face; the face that had been hovering over her. Mac was not in the meager cabin.

"You must be getting pretty hungry by now."

Laura didn't comment. It was all she could do not to retch as the bile rose in her throat.

"How's the bed? Nice and cuddly, isn't it? Ya know, it takes me months to get enough of those little hand warmer packs up here to heat the bunk. The more you move around the warmer you'll feel. That's why they're so perfect for fuckin'. If only I could market the concept. It's been a long time since I've entertained a lady up here."

I can see why.

"I guess most females don't like the long walk." He snickered at his own remark. "Course, they never complain about the walk out."

"If you treat them like this, I'm sure they never make it out." Laura was angry at herself for snapping at the man. She didn't want to give him the satisfaction of carrying on a conversation with her.

"Well, I guess you gotta point there. It always seemed like too much work to take 'em out, if ya ask me. Once I've had the satisfaction I want, it always seemed easier to be done with her. If you get my drift." He walked up to her with a cup in his hand. "Drink this; it'll make you feel better."

Laura turned her head away. If he was going to kill her, she didn't need to cooperate. On the other hand, cooperation might buy her some time. Somebody had to be looking for her. Where was Mac? He had to be around here somewhere. Then it dawned on her. Mac wasn't going to be seen anymore. He wasn't part of Jake's party; and, most likely, Jake had made sure Mac wasn't walking out either. "I don't want anything."

"You'll drink this." Jake place his thumb and forefinger on her jaw pulling it toward him. Squeezing until the power of his grip forced her to open her mouth, he slowly poured some of the warm liquid into her mouth. Before she could spit it back at him he shoved her mouth closed and waited for her to swallow.

"Why do you do this?" She snapped after feeling the warmth of the liquid.

"Hell, I'm a man. But then you know that by now." He grinned at her with yellowing teeth showing through the long whiskers. "I even got paid for this job. Usually, I have to go all the way to Denver to find some gal, who won't be missed, and drag her up here."

"Why don't you just pay for a prostitute?"

"Hell, that's no fun. I'm a sportin' man. This is more fun."

"Having your prey tied to a bed isn't exactly a sport."

"It's all in the eyes of the beholder, ma'am. Besides, this mountain's my home. I know every bit of it; and if I need a quick get away, I'm better off here." Jake took off his shirt revealing the huge scare on his left shoulder. He watched her eyes as she stared at him. "Yeah, this one nearly killed me," he brushed his finger over the white scar tissue. "I've got a few others, wanna see?"

Laura turned her head away from him. She was feeling tired again. Beginning to drift. Her eyelids were like logs wanting to drift down stream. She couldn't control the current. *The son-of-a-bitch drugged me again.*

She heard his husky voice as he towered over her. "That's right close your eyes. You'll not sleep so soundly this time. It'll be more sportin'." The grin was broad and possessive.

Laura tried to force her eyes open one more time. She had to focus just for an instant. The lips came together while her tongue pushed the saliva up to the front of her mouth. When her eyes opened, she spit in his face. His laugh haunted her as she drifted on the edge of consciousness. She couldn't hide in the bed or in her mind from the hands touching her body.

****

"Megan, all I could see through the window was one man walking around. I didn't see your mother. I told Gramps to put in bigger windows, but he'd always tell me to go outside if I wanted more light."

"What do you suggest?"

"There's a fire going. It's easy to climb onto the roof. I thought I'd shove something down the smoke stack. The cabin'll fill with smoke. Eventually, one of them has to come out and see what's wrong."

"Sounds good to me, but what if this guy isn't one of the kidnappers?"

"I don't really care. He's trespassing on my land. I know I haven't been up here for a while, but it's still my land. In fact, we never even knew how Gramps came by ownership of all this. We figured it was through some federal land grant for mineral rights. It didn't matter. He had the deed which was passed to my father and then to me. The man, whoever he is, is in my cabin; and right now I want it back. Besides the smoke won't hurt him or your mom if she's in there, but it'll certainly flush him out."

"Let's take a look around the place first. Maybe we can find something to overpower him with when he does come out."

"There's not much to look at. A wood pile is in the lean-to on the other side. Let's see if there's a good size log I can wield at his head."

Megan looked at everything as Luke led her around the edge of the cabin. The sun, and the limited light it offered, was going down past the peak of the mountain, making it even more difficult to differentiate ground from sky. She would have taken her helmet off in order to see better, but then she'd have to raise her voice to talk to Luke. He had been right; there wasn't much to see except white. Anything Luke might have used as a weapon was buried under several inches of snow. She could touch the cabin now as they turned to follow the back wall to the lean-to. Only the howl of the wind was heard by those inside and outside the cabin.

Luke reached the lean-to first and motioned for Megan to stay behind him. He nearly disappeared as he moved inside the dilapidated wooden shed. When he turned around, he carried a log about the size of his arm in one hand and a hefty size knife in the other.

"Where'd you find that?" She pointed at the knife.

"Off the dead body shoved inside."

"A body? Any idea who?"

"Never seen him before, but he kind of matches the description Brad gave us about the man following your mom. We know one thing for sure." Luke paused waiting for Megan to absorb the seriousness of the situation. "The man inside is dangerous."

"No shit! I could figure that part out for myself. If I wasn't trying to save my mother from this animal, you can bet I'd be hauling my carcass out of here a hell of a lot faster than I came in. Why do you suppose 'jerk number one' didn't take the knife before dumping 'jerk number two' out here?"

Luke liked her sense of humor. It eased his tension and increased his confidence in Megan's ability to handle the situation. "'Jerk number one' didn't need to worry about it. This guy isn't going anywhere, and there's probably plenty of guns and knives inside."

"What are you going to put in the smoke stack?"

"I was thinking about pine branches now that I can use this to cut down the bigger ones." Luke held the knife out in front of him. "Go to the corner of the cabin. If anybody comes out the door, tell me. I'm going to grab a couple of branches."

Megan did as she was told. She didn't want Luke to fight the man, but she didn't have any other suggestions to make. This guy was obviously very talented at his evil games, and she didn't know if Luke would be able to match him in strength or wit. After all, Luke was a businessman and part time rescue member, not a fighter. Glad to be wearing the headset so they could still talk to each other without being heard inside the cabin, Megan expressed her fears to Luke.

"Luke, I don't like you trying to take this guy on alone, especially if he's not the one we're looking for."

Steel Illusions

"He's the one we're looking for. The I.D. I found on the body had a Tulsa address. That can't be a coincidence."
"You're right. "Did you get a name?"

"I looked, but I don't remember. It was the city which caught my attention."
"Maybe 'jerk number one' already killed Mom. Luke, I don't want you to risk your life if there's no reason to."
"Are you worried about me?"
"Yes." She said flatly.
"I appreciate the concern, but just because I didn't see your Mom, doesn't mean she wasn't in there. I couldn't see the bunk at all."
"Just make sure you knock the shit out of the guy when he comes out of the cabin."
"I'll give it my best shot."
"Luke, what should I do if you're hurt."
"Run like a rabbit." Luke was standing beside her when he gave the last piece of advice. He'd rather take off his helmet and give her a big kiss; it might be the only chance he ever had. "You watch the door again while I get onto the roof. It's easy access from the back. If I make to much noise, he'll come out before I'm in position."
"Just don't fall off the roof. I don't think I can carry you back up the mountain."
"I don't think you'd have to worry. We'd both be dead." Luke took his helmet off handing it to Megan. He needed better vision and easier movement of his neck and head.
"Thanks for the optimistic words." She said as he walked away from her. Megan didn't need her helmet any longer since she couldn't communicate with Luke. She set both of them on the ground by her feet and waited.

CHAPTER FIFTEEN

Amy closed the door as the last guest finally left. She'd had company constantly since Pete had died. Her anger seethed just below the surface of her crumbling facade. The noise in the kitchen grated on her raw nerves, so she walked into the bedroom closing the door behind her.

Warm shower water ran over Amy's face doing a minimal job of comforting her outer aches and pains. She turned off the water and wrapped the large white towel around her thin body. There was still no visible sign of a baby. Her wet hair dripped down her back as she yanked the comb through the tangles and stared at her reflection in the mirror. The face was pale, and large gray circles framed her eyes. There were no more tears, only anger.

The light knock at the door wasn't enough to make her draw away from the sad reflection. The woman standing draped in a white towel was not the person Amy remembered being. Before Molly's illness, she'd been so happy just living a simple life and loving Pete. Then there were the long months of hospital life with Molly. The smell of death which had permeated the room the last days of her life all came back to her when she'd been in the room with Pete. A lifetime of sorrow had been lived in the long days just after Molly died; and, now she was reliving the hell alone. At least after Molly had died, Pete had been at her side. They had nurtured each other through the difficult days and months of grieving. Now only her anger kept her company; only her anger got her out of bed this morning.

Again the light knock at the door. "Amy, may I come in?"
"I don't want any tea tonight, Mother. Go home."
"I have to talk to you."

Steel Illusions

"I'm talked out. I don't want to hear any more trivial clichés about time healing my scars; and someday, I'll find someone else. The arrangements have been made for tomorrow; now just go home."

"I won't go anywhere until I talk to you face to face. Now do you want this cup of tea or not."

"Fine, come in." Amy's irritation was evident in her curt invitation.

Amy's mother opened the door and walked in confidently. This was not the time to be hesitant. Amy needed to hear part of the truth, and the mother could be just as stubborn as her daughter. She put the cup of tea on the night stand and moved to face her daughter.

"I know you're angry. Part of that's normal, but I think you're letting your grief manifest itself into revenge."

"I don't know what you're talking about."

"Yes you do. I saw you talking to that man again. The one who promised to help you get even with Kevin Bradford."

"If Bradford is guilty, I have a right to want revenge."

"Listen to yourself. Revenge? Pete's concern was always for you. Now you have to think about the baby. You owe it to him."

"He's not here anymore. And my concern is for this baby. Pete was always worried about money; now I'm going to make sure we don't have to worry anymore."

"By suing the company? You can't sue Bradford or the company. Even if you do file, you can't hurt Bradford directly. Take the settlement the company offers you and walk away."

"Not a chance, Mother. This is not for discussion any longer." Amy tried to dismiss the woman standing in front of her, but the feisty, older lady didn't move.

"Amy, you're not thinking straight."

Amy's face was red with anger. "I don't care!" She shouted and threw her cup of tea across the room. "Get out!"

R. Z. Crompton

"Scream at me if it'll help. But you will hear me out."

"You don't know what it's like. How could you? Daddy was with you for over twenty years. You didn't have to raise a child alone."

"Have you become so insensitive?" She shook her head at the young woman. "Your father's sudden death was as hard on me as this is for you. I was angry too, but I had you and your brother to focus on. If you let your anger fester, you'll become a bitter old woman with no life, no hope; and you'll pass that bitterness to your child. Is that what you want?"

"You don't know what you're talking about."

"Yes I do. I've seen it happen many times. I watched my own mother shrivel up and die after my dad was killed in a car accident." She finally had Amy's attention.

Amy remembered back to her childhood. She used to sit on the lap of a kind old woman with long silky gray hair which she always wore in a tight little bun at the nape of her neck. Granny had told her funny stories and sang to her when she'd been sick. "I remember Granny being soft and cuddly. When did she become bitter? Why didn't you ever say anything?" Amy sounded skeptical almost like she didn't believe her mother.

"There was nothing to say. My mother died when I was very young. My grandmother raised me. That's who you grew up calling Granny. You sat on my grandmother's lap. I never talked about my mother because all I remembered was how angry and sad she was for the few years I had her. Amy, time doesn't heal all wounds. Only you and God can heal you."

"God? How can you talk to me about God? The God I believed in wouldn't have let Pete die so painfully." Amy walked across the room wringing her hands together as if she could strike at this God she was trying to hate.

Steel Illusions

"Amy, God didn't let Pete die; He released him from his pain. Pain, I believe, he caused himself."

"How can you say such a thing? Leave, Mother. Just leave me alone."

"I'm going to say it, and you're going to listen. Tell me why Pete went down onto the furnace floor without his flame retardant coat on. Can you tell me?"

"No. Maybe he forgot."

"Forgot? I don't think so. How could he forget something he'd worn everyday for four years. I think he did it on purpose, so you would get the money from the insurance and the company settlement."

"Impossible. I won't believe it. I can't."

"You'd better, Young Lady, because if you pursue a lawsuit and the company proves Pete's actions were intentional, you won't get a dime from the insurance company or the steel mill."

Amy didn't say a word. Her mind was reeling with possibilities. Thinking of Pete facing the molten steel waiting to be burned alive brought the latent tears closer to the surface. She thought again about the last words he said to her. A letter? He'd mentioned a letter.

"Mom, what if only part of what you're saying is true?"

They were making progress. At least her daughter was thinking about other possibilities, and she'd had gone from using a very sterile "Mother" back to the more loving "Mom".

Amy sat back down and picked up her brush. Her mother had seen the ritual ever since Amy'd been a little girl. She'd sit for hours brushing her long blond hair and thinking. The stroking was a cleansing ritual; it helped her think clearly. "Remember the man at the hospital?"

"Yes, he was here today. I really dislike the guy. He's greasy, like he's been dipped in slime. He's the one who promised to help you. I wouldn't trust him to bring me a drink of water."

R. Z. Crompton

"I feel the same way, but I wanted to hang Kevin Bradford so badly, I didn't pay attention to what the man was really saying. Just before Pete died he told me there was a letter and to use the money for the baby. I didn't think much about it, but Mr. Waterman keeps asking if Pete left a night letter at home, so it could be used as evidence."

"I'm not following you, Dear."

"Pete never brought home a night letter. I heard him talk about them, but I never saw one. It was just a list of instructions which he would throw away when he was done. Why would Pete tell me there's a letter unless it was something special. A regular night letter wouldn't be anything special. Nothing to tell me about. Maybe Pete was part of a plan. Maybe the explosion wasn't an accident."

"Have you checked the mail?"

"Why? It's always junk mail and bills. I didn't want to deal with any bills just yet."

"Pete said there was a letter. Maybe he was talking about a real letter."

"I haven't opened the mail box for three days." Amy pulled on her bathrobe and headed for the front door. She didn't know if she wanted to find anything or not. The small, black box, attached to the wall by the front door, was full of advertisements and the normal white envelopes. She flipped through the rectangles, recognizing the majority of the return addresses. One envelope caught her attention; the hand writing belonged to Pete.

"Mom, maybe this is what we're looking for." She handed the suspicious white paper to her mother. "You open it. I can't be as objective as you can."

Her mother was quiet as she read the two sheets of paper which had been neatly folded and placed in the envelope. The first page she perused was obviously the night letter with Kevin Bradford's initials on the bottom. She quickly scanned the second page and handed it to her daughter.

Amy slowly read Pete's profession of love and his explanation of the plan he'd devised to take care of their money problems. The tears filled her eyes and flowed down her face. Her anger was gone and sadness filled the endless space. Now it was a matter of honor. Giving this stuff to Kevin Bradford would certainly let him and the company off the hook; unfortunately, it would also mean Pete had died in vain because there would be no money.

"What am I going to do?"

"For now, remember why your husband died and how much he loved you. Go to the visitation tomorrow with the support of the people who love you. You can make decisions after the funeral on Monday."

Amy's shoulder's heaved with the sobs she'd not shed all day. Pete had only been dead for twenty-four hours. And the short amount of time had been filled with enough anger to last forever. Her mother held her tightly until, gradually, the tears subsided. Amy didn't know, or care, if the tears were for Pete or merely a cleansing of the rage she'd felt. Her fury should be directed at Pete, but there was no energy left.

"Mom, why wasn't I mad at Pete?"

"You didn't want to be. But you will be from time to time. It's normal to be angry at the one who died leaving you alone to cope with the world. I've spent many days being mad at your father. It passes."

As Amy dried her red eyes, her mother added, "Please stay away from Mr. Waterman. He's no good. In fact, I'd feel a lot better if you came to stay with me until all this settles down. I wouldn't want that guy to show up here with you all alone."

"I'd like to be brave and say you're silly, but I think you're right. This afternoon, he kept touching me. I pulled away from his hand over and over, but he was so close. He asked me when we could talk alone. The way the words slid off his tongue, I didn't get the impression he wanted to talk at all. I'll get dressed. Mom, thanks for being so stubborn. I really need you." Amy walked over and hugged her mother.

"I love you too. Now hurry up."

****

"Good evening, Kevin. How are you?" The manager of the Warren Duck Club greeted his good customer and friend with the casual address he saved for only a favored few. His job required the more traditional formality.

Kevin received the warm handshake and greeting with a smile of his own. The good nature of his host was infectious; and especially tonight, Kevin appreciated the friendship. "Hello, Alex. I see you have a full house tonight. Do you have a quiet corner for me." Kevin looked around the elegant restaurant. The soft lighting accented pastel colors and deep rich wood. Laura's favorite place to sit was near the wall of windows. The copious view of the trees and gardens below was decorated with the tall church steeple just above the tree line. Normally, the setting sun offered a rainbow of colors to dine by. Kevin didn't want to sit by the windows tonight.

"Absolutely. I saved a special place for you and your guest." Alex led Kevin around the small palm tree into a corner of the restaurant. "I heard the terrible news about Laura last night. Anything I can do for you?"

"No. Thanks for asking. I know she'll want a wonderful homecoming celebration."

"You will have it here, please?"

"Naturally, Alex."

Steel Illusions

"You've had your own troubles at the mill, too. I'm so sorry." Alex's heavy accent was pleasing to the ear after the harsh sounds Kevin had been hearing all day. It would be nice to be pampered with excellent service and food.

"What did you hear on the news about Laura? I didn't get to the see broadcast myself."

"Just that she was missing and there was a delay in the search because of inclement weather. What about your beautiful daughters? Nothing was said about them. I thought they went with Laura."

"They did. In fact, they are out in the wilderness looking for her."

"Oh, my God. They are so brave."

Kevin smiled at Alex's remark. He knew why the man was so good at his job. Alex was genuinely concerned about every person who came into his elegant restaurant. The service and atmosphere were the best, but Alex, with his warmth and charm, took the restaurant to another level.

Rick was escorted to the table as soon as he came in. He looked as intense as Kevin felt. He quickly ordered a glass of red wine and placed his cellular phone on the table beside Kevin's beeper.

"Are you expecting a call?"

"Not really. But you never know. I see you finally made it home to retrieve your little, black box."

"I wanted to see if the girls had left a message for me. There were several, some from Laura before the kidnapping and some from the girls desperate for me to call them. Even my beeper had several numbers from the Beaver Creek area. I'll never walk away from that damn answering machine again without checking the messages."

"Did you call the rescue office today?"

"Yeah, this morning. There was no news. The girls spent the night in a snow cave with the other members of the team. Several people died in the avalanche. Today's Friday, they were supposed to be home tonight. Krista wanted to spend a couple days with me before going back to school. Now I don't know if I'll ever see any of them again." Kevin gulped down his glass of wine hoping to curb his rising emotions.

Rick changed the subject. "You said you'd found something."

Kevin cleared his throat before answering. "I spent the afternoon comparing the financial reports for the last five years with the scrap records. You were right and wrong. The big scam wasn't only the nightly removal of scrap we'd already paid for. It was the annual adjustment that's been made for at least the last five years."

"Explain please."

"I'll try. The annual physical inventory shows that each year the meltshop was short as much as five thousand tons. The physical inventory is compared to the records of incoming scrap shipments and how much of that scrap was used in making steel. We can be high on the inventory or low."

"So how does that end up in somebody's pocket? It doesn't seem like enough to kill for."

"That alone isn't. The company took the hit every year for the adjustment. First of all, the variance shouldn't be that great. Second, the variances should average out; with some years short and others long on scrap estimate."

"I still don't follow you." Rick stated as the waiter came to take their order.

Kevin was quick to order his favorite: the Blackened Thick Tenderloin, and his mind shifted to Krista, who shared his opinion about the perfectly seasoned piece of meat. "Rick, it's the best cut of beef this side of the Mississippi."

"I thought you were sticking to fish these days."

Steel Illusions

"This is one of my pleasures in life, and tonight I'm going to indulge."

Rick didn't blame his friend and ordered the same before asking him to continue his explanation.

"Every month we estimate how much scrap is in our inventory out back. It's an asset just like the paint or lumber a carpenter uses. The excess we get credit for on the asset ledger; the shortages we have to deduct. For example, a carpenter uses two cans of paint instead of the three he bought. He'll credit his own records with one gallon of paint. It's the same for us. We don't always use as much as we estimate. Sometimes we use more. Since I've worked here, I've tried to cut back on the physical inventory because it's so hard to calculate. What really bothers me is that before I came, scrap was constantly being stock piled. How could the physical inventory be so short at the end of the year? The annual physical inventory is audited which means something is fishy with the monthly estimates, which aren't audited."

"In other words, what you basically have on the ground doesn't match what's in the computer. Has the annual physical inventory off during the last three years while you've been here?"

"Yeah, it's off, but not as much."

"So you've curtailed some of their activity, and they don't like it. What about all the tracks out by the scrap piles?"

"I think somebody's been trucking scrap in and out of the shop for years, but a truck load here and there wasn't even noticeable, especially if it was done when the weather would cover the tracks."

"Or when there wouldn't be any tracks left at all. Sometimes it's so dry, a little breeze would get rid of any sign a truck had been around. There was no hurry. They could bide their time 'til conditions were right and then haul like crazy."

"That's about it. I think the explosion is only one part of a plan to get rid of me. I'm just not sure yet how or who is planning to do it."

The waiter delivered the meal as Kevin and Rick discussed the possibilities. Rick was the one to deliver the warning Kevin hadn't thought of. "These people are very dangerous. They've already killed here and in Colorado. I'm afraid you may be on the list."

Kevin nearly choked on his beef as he realized the strong possibility of having his name on a "to kill" list.

"I may have another piece of information for you by Monday." Rick stated while he waited for Kevin to digest the likelihood of being a marked man.

"What's that?"

"I called in a favor from one of my cohorts on the east coast. The pieces of burnt scrap you wanted analyzed weren't even scheduled to be done for about two weeks. My buddy is going to go into the lab and run the tests over the weekend. He'll call me as soon as he finds anything."

"Do you really think it'll help."

"Absolutely...." Rick's comments were interrupted by the beeping of Kevin's pager.

Kevin picked up the little black box and read the number. It was from Colorado. "May I use your phone?"

Kevin punched in the numbers and waited for an answer.

"Hello." Kevin responded to the answer with his own identification.

"Mr. Bradford, thank you for calling back so quickly. My name is Tim; I'm in charge of the rescue mission."

"Do you have any news about my wife?"

"No, I'm sorry. I did want to let you know I talked with Krista this afternoon. The rescue team was caught in an avalanche just after lunch."

"Oh, my God. Are they okay?"

"Krista was slightly injured in the slide, but she's feeling better now. As soon as we can get her off the mountain, we'll have her taken to the hospital. Her two friends were more seriously injured, but I think they'll be okay. A local doctor was on the radio with Krista shortly after the incident. Medical supplies were dropped to them about thirty minutes later."

"Meg? What about Megan?"

"She's fine as far as I know. She and Luke, one of our rescue members, traveled on. They were only a short distance from the cabin which we suspect is the hideout for the kidnappers."

"Was the avalanche man-made?"

"We don't think so. From the description given to me by the rescue member staying in the cave with the injured, we believe it was a natural slide."

"What about the weather?"

"The forecast calls for a clearing sometime tonight. We should be able to fly the injured out tomorrow."

"If you flew supplies in, why can't you get the kids out."

"We had a plane drop the supplies, but even a drop was risky in the high winds. The plane can't land in the mountains. Both of the choppers we have emergency access to were flying the I-70 avalanche victims to Denver. Besides, visibility has been so poor we couldn't see anything, let alone fly a rescue. We'll get a chopper up there in the morning."

"What about the search for my wife?"

"The places we could check from the highway have been searched. The stocked huts in the mountains will be checked tomorrow. We really don't think the kidnappers would take Mrs. Bradford to such an obvious location, but we're looking anyway. Once the weather clears, they won't be able to move around without us finding them."

"You sound very sure of yourself."

"I'm good at what I do, Sir. The people who work with me are very good. Road blocks were set up right away; and with the bad weather conditions, travel was nearly impossible. Your wife was not taken out of this valley. Luke and Megan will be checking the most obvious spot as we speak. That's why they left so quickly after the abduction. The idea was to surprise the kidnappers before they'd be watching for us."

"How can two kids overtake dangerous criminals?"

"Excuse me, Sir. I appreciate the fact that these are your daughters; however, they're not kids. They are strong, young adults; and they're with rescue people who know what to do."

"I am the absent father, and I'm concerned. My wife could be dead, and both my daughters were in an avalanche while I'm stuck here at a damn steel mill. The balance is wrong."

"I understand your concern and your absence. Any idea who might have done this?"

"Ideas, but nothing concrete. Can I be of any help out there?"

"Not really. We'll call as soon as we pick the girls up. I agree that whoever planned this wanted you out here on the first available plane. You and your family won't be safe until the person or persons are caught. We'll take care of things out here; you try to find the bad guys."

"Thank you." Kevin cut the conversation because his pager had identified a call from the mill. "May I?" He motioned at the phone again, and Rick shook his head.

Steel Illusions

"Aren't you glad I brought it. Why don't you join the age of technology and get one of your own."

"I don't need one. I can use yours." Kevin's smile was shortened by the answer of the phone. Dan's gruff voice was at the other end of the line.

"Hey, Boss. I found something really weird."

"What?"

"I saw Waterman coming outta your office a little bit ago; made me curious. Ya know? What'd he be doing in there so late?"

"Did you go in and look around?"

"Of course I did. Being nosey is part of my job. Don't you remember seeing it in the job description." Dan's remark brought a smile to Kevin's face. The man had the biggest eyes and ears in the shop.

"I didn't think you'd read your job description. What'd you find?"

"Some papers sticking out from under your desk pad. They weren't there earlier."

"What kind of papers?"

"It's a scrap report for the last several days."

"So what's wrong?"

"The numbers are screwed-up."

"Dan, did Waterman see you?"

"Don't think so. Why?"

"If he's the one who set all this up, he's dangerous. I'll be right there. Don't leave the mill and don't get caught alone."

"Sure thing, Boss." Dan's voice was filled with the same, jovial gusto it always held. He didn't appreciate the danger.

"Dan, I'm serious. If he thinks you're in his way, he'll try to kill you. I'll be there as soon as I can." Kevin turned off the phone and got up to leave. "Finish your dinner. I gotta go."

R. Z. Crompton

"No way. I'm going with you." Before Kevin could protest, Rick had handed his credit card to the waiter asking him to hurry.

"Should we call the police?"

"Yes, but not the blazing lights. I'll call...."

"A friend?"

"Somebody I know who won't laugh at us if we call him out on a wild goose chase. I'll follow you."

CHAPTER SIXTEEN

Smoke quickly began to fill the small cabin. Jake threw the warmth of the animal furs off as he reluctantly jumped out of the bunk. "What in the hell! Snow must be clogging the smoke stack. Shit! Damn cold weather." He grumbled to himself pulling on his pants and coat.

Laura moaned as he spoke. She didn't understand or care about his absence. Her semi-conscious state only permitted her enough awareness to realize he was gone, allowing her a few moments of peace.

Jake's first impression as he opened the cabin door was that the snow had stopped. *Funny, damn thing shouldn't back-up if it's not snowing.* His peripheral caught the movement and instinct caused him to thrust forward taking the blow on his shoulder rather than his head. He was stunned but not incapacitated.

Megan's breath caught in her throat as she watched him stand up and square off to face Luke. This is what she'd feared most: the two men fighting one-on-one. She was certain Luke would be at a disadvantage. She doubted he'd ever had to fight for his life before, and this man was obviously very experienced.

Before Luke could respond, Jake pulled a long knife from his coat pocket. He lunged at Luke who barely managed to get out of the way of the first slash. Megan gasped causing Jake to intuitively look her way. The instant was enough for Luke to get his balance and swing the log cracking the kidnapper in the chest and knocking him over.

Jake scrambled to his feet. A dozen years living in these mountains had made him dangerous and agile. It'd take more than a log to defeat him. He squared off to face his attacker. The face was

Steel Illusions

familiar, but Jake wasn't sure where he'd seen this young man before. It didn't matter. He was the enemy now, and he was about to die.

Luke didn't dare drop the log. It was all he had to offer some self defense. The knife he'd found on the dead man was in his belt, but he hadn't thought to take it out of the sheath. The seconds required for him to free the knife would be all this guy needed to turn him into a shish kebab. The man came at him again. Luke tried to get out of the way, but the long curved knife caught it's target. Luke fell back grabbing his upper arm. Pain burned through his entire body as the acknowledgement of being cut reached his brain.

Megan was desperate. *If he kills Luke, he'll come after me for sure. I won't stand a chance out here against him.* She had nothing for defense, only the two helmets. *What in the hell am I going to do with two helmets? Luke's going to have to start carrying a gun if I'm ever going to do this with him again.* Images of Indiana Jones played through her mind as she saw the hero pull out a gun shooting the man with the sword. *Shit! A gun would've come in handy about now. Well, I've got no choice. I can't let the bastard kill the man I'm going to marry.*

Luke, watching the predator take slow pronounced steps toward him, tried to scramble to his feet again. Their eyes locked and Luke waited for the man to surge forward again. It was obvious the killer was enjoying the stalk as a slight grin appeared. Neither man noticed the slight movement from the side of the cabin. A helmet came flying toward them causing the killer to turn away from Luke just for an instance. It was all the time Luke needed to wield the log at the killer's head dropping him first to his knees and then face first in the snow.

Megan ran to Luke and threw her arms around him. "Are you hurt badly?"

"Ouch!" Luke winced at the pain. "My arm hurts like hell, but I'll live." He wanted to kiss the woman who'd saved his life and her own, but this wasn't the time. "Thanks, Meg. I never thought of a helmet as a weapon before, but I'm certainly glad you didn't feel limited."

"It's all I had, and I couldn't let him kill the man..." Megan hadn't even been aware of her marriage thoughts until now. She'd nearly stuck her foot in her mouth by confessing her thoughts to him. She stumbled on her words before blurting out "...the man who was going to save my mother." Nice recovery, she hoped as she turned and looked at the body on the ground.

"Stay back. I don't know if I killed the son-of-a-bitch or not." Luke tentatively tried to see if the man was still breathing. Blood was oozing from the back of his head and there was not a noticeable pulse. "I think he's dead. At least he can't hurt us anymore. Let's go find your mother."

Megan was afraid to go into the cabin. She didn't know what she'd find. Her movement was filled with reservations as she stepped toward the door. Luke sensed her trepidation and asked, "Do you want me to go first?"

Megan didn't answer, only moved aside to let him pass in front of her. Holding his arm tightly, trying to curb the flow of blood, Luke walked into the cabin. At first, he didn't see anyone and his heart sank thinking their trip had been in vain, then he heard a moan from the bunk. If his suspicions were correct, it was Laura; but he didn't know if he should let Megan find her tied to the bed or not.

"Meg, your mother is alive, but I'm not sure you're going to want to see her....."

"Why? What's wrong?" Megan had horrible vision of the man's knife utilizing her mother as a carving post.

Luke was facing her just beyond the threshold of the cabin. "I'm not sure, but my guess is this man..." his words were full of disgust. *What were the right words?* Luke hesitated.

"Did he cut her?"

"I don't know; she's in the bunk under some furs. I think..."

"I understand. Better me than you finding her this way." She walked past him acting more as a rescue member than a daughter.

Megan cautiously approached the bunk nearly jumping out of her skin when she heard the deep moan. The wrists tied to the top of the bunk were bloody and raw from the rope. Megan was sure the ankles would look the same. *What kind of sick son-of-a-bitch would do this?*

"Mom, it's me. Mom?" There was no intelligible response only another moan.

"Luke, cut the ropes. The least we can do is make her comfortable." Luke did as he was instructed, but Laura started to struggle as soon as she felt the pressure on the ropes. Luke barely missed cutting her hand.

"St....aaaaay awaaaaay.....frooooommmm mmmmme!" The words were slurred and difficult to understand.

"She's been drugged." Luke's anger rose. He wanted to castrate the man who'd done this.

"Mother, it's Megan." Megan reached over and tenderly took her mother's face in her hands.

Laura tried to struggle, but Megan held her firmly turning her mother's face toward her. "Mother, open your eyes and look at me." Again she didn't let her mother pull away but lovingly controlled the movement with one hand and stroked Laura's forehead with the other. "Mom, open your eyes. It's all over."

Luke was nearly in tears as he watched the love pass from daughter to mother. Megan, with a soft caress, was passing the will to live to her mother. No wonder she was so good with animals. She could do with the sound of her voice and tenderness of touch more than anyone he'd ever seen.

R. Z. Crompton

Laura responded to the familiar voice by opening her eyes and trying to focus on the two people standing over her. Recognition was slow and tentative. "Meeggaannn?"

"Yes, Mom. Do you understand me?"

There was no verbal answer, only a slight movement of her head.

"This is Luke. Remember?" Megan spoke slow enunciating every syllable. "He is going to cut you free. Be still. Okay?"

Again Laura gave a slight affirmative nod.

Luke freed her other arm and her legs, but she was still too groggy to understand she was free. Megan brought a cup of water, but Laura frantically pulled away from her shaking her head.

"No! No more."

"Meg, he must'a drugged the water. Calm her down, and I'll see what else I can find. By the way, don't you drink the water either. I don't need you jumping around like a scared rabbit. We've had enough trouble with this relationship."

Megan was surprised to hear Luke think they even had a relationship. Maybe there was hope for him. She felt guilty when she looked up at him and saw the blood dripping down his arm and onto the floor. "Luke, we've got to stop the bleeding."

"Don't worry about it. I'd forgotten about it myself."

"You won't be any good to me if you bleed to death. Now take your snowsuit off, so I can take a look."

"Yes, Sir. Doctor, Sir."

"It's Ma'am. Doctor, Ma'am. And that's better. I'll calm Mom down and then take care of your arm. Look and see if you can find anything I can use for a disinfectant."

"Yes, Ma'am." Luke smiled as he proceeded to do as he was told.

Steel Illusions

Laura was curled up in a ball on the far side of the bed. She still didn't realize the danger was over.

"Mother, come on." Megan coaxed, hoping she could convince her mother to relax and rest. They wouldn't be able to do much for her until the drug wore off. The best they could do was keep her warm and comfortable for awhile.

With soft reassuring words, Megan finally had Laura resting peacefully in the warm bed. She then turned her attention to Luke. He'd taken his snowsuit off to his waist, then his sweater and undershirt. Megan's attention had been on her mother as Luke undressed; so when she turned around, she was looking at an extremely muscular man with broad shoulders and a sexy, hair covered chest. In spite of the pleasure she experienced while evaluating his handsome features, Megan was concerned about his peaked complexion.

"Here, sit down and let me look at your arm." Megan walked over to the man sitting at the table with a bottle of whiskey in front of him. "Is that going to help you internally or externally?"

"Both, I hope. It was all I could find. I don't think the man would've spiked his own whiskey. At least he had an appreciation for the good stuff."

Megan looked at the square bottle of Jack Daniels sitting on the table. "Did you wash out a glass?"

"Actually, I was thinking it'd just be easier and safer to drink out of the bottle."

"I'll get us both a clean glass after I'm done looking at your arm. Did you take a couple swigs or are you fading because of the pain."

"A little of each. I should bring in more fire wood. It's getting cold in here."

"That can wait for a few minutes." Megan took her first good look at Luke's arm. The clotting was good, but she was concerned about infection. Pouring whiskey right out of the bottle onto an open wound reminded her of tales from the old west. For now it was all she had. *I hope those old movies were being authentic. I never read about this in any of my medical books.* Before pouring the Jack Daniels on Luke, she took a big gulp. *Shit!* Megan coughed as the whiskey burned its way down her throat. "I don't know if this will burn your arm as much as it just burnt me. Are you ready?"

"I guess so. Have you ever done this before?"

"Nope. Only watched it on the movies. Frankly, we're a little more civilized in Texas."

"Just do it, Smart Ass."

"Don't get testy with me. You're at my mercy."

Luke leaned over and kissed her before she could pull away. "That was a mercy kiss, now pour before I change my mind."

Megan burned again but this time from the heat of Luke's lips on hers. She felt dazed as she tipped the bottle allowing the brown liquid to wash down his exquisitely defined biceps.

Luke's jaw tightened as the liquor burned into his flesh. His fist clinched the side of the table while he tried to control his urge to yell a string of four letter words he'd heard only from his grandfather when Luke had accidently dropped the man's favorite rifle off the top of a cliff. It was the first time he'd yielded to the memories of his best friend. As he looked into Megan's tear filled eyes, he believed he'd found a new companion.

"Are you going to live?" She asked tenderly.

Luke answered with an affirmative nod and closed his eyes until the burning sensation had passed. Not willing to wait for a clean glass, Luke took three big gulps from the bottle. He watched Megan rummage around the cabin looking for something, he assumed, she could use as a bandage. She needed help, but he could hardly move his good arm.

Steel Illusions

"I found this medical box stuck back in the corner. It's full of gauze and aspirin. You need about twenty stitches, Luke, or you're going to have a hell of a scar. I'll wrap you tight as I can. I'm really sorry, but that's about all I can do for you."

"Don't worry, Meg." His words were beginning to slur more from the Jack Daniels than the wound. She imagined his head would hurt more than his arm in the morning. He'd drank nearly half the bottle in less than fifteen minutes, except for what she'd poured down his arm. "You saved my life, Meg."

Smiling, Megan looked at him. When she realized how drunk he really was, she felt irritated with him. She'd hoped they might share a nice evening together while her mother slept. *Oh well, it's better he sleeps too. If he was awake, he'd just want to help; and that wouldn't be good for his arm.*

Bed? She needed some place to lay him down, but she didn't think her mother would be able to deal with having a man in her bed. Megan pulled some of the furs off the bunk and made a bed for Luke on the floor.

"C'mon, Sweety." Megan tried to hoist Luke onto his feet.

"Sweety," he slurred. "I like the sound."

"Yes, I like the sound too." Megan tried to humor him as she moved him to the furs on the floor. It wasn't going to be a very soft bed, but then he wasn't going to notice.

"I like you too, Meg." He had a big grin on his face as he sunk down onto the floor. "Will you marry me, Meg?"

"Sure, in the morning."

"Promise?" He sounded like a little boy.

"Yes, I promise." She smiled then added, "Now go to sleep." Megan was alone and disappointed. If he'd been sober, he never would have said those things. At least he asked while he was conscious. That was better than nothing. Not much, but still better.

Hell, it might be the only proposal she ever got. Luke was snoring before she had him covered with another fur.

"I guess it's up to me to bring in more firewood before it gets too dark." Megan spoke out loud. It was better than the silence, and there was no one to care if she talked to herself.

Megan checked to make sure her mother was sleeping peacefully and then zipped up her snowsuit to brave the still frigid wind. She closed the door behind her and headed for the lean-to. An arm load of wood wouldn't last long. She'd have to make several trips in order to have enough for the entire night. After two trips she found herself tired and winded. It'd feel good to be finished with this job. Finding something to eat would be much easier. She was starving.

Megan carried the third load into the cabin and froze in her steps. There should've been a body in the wood pile. There should've been a body outside the cabin door. Both bodies were gone. Megan dropped the wood at her feet and tentatively looked out the cabin door. Adrenaline rushed through her body giving her a nauseous feeling in her stomach. Nothing out of the ordinary came to her eye. *What's wrong with this picture?* She asked herself the question several times and each time she gave herself the same answer. *There should be a body and there isn't.* Megan slowly walked to the location she remembered as the "downing of a killer". The spot was bare. Not even a foot print was visible. The wind was blowing enough to cover the foot prints but not the body. There was only one explanation. The man wasn't dead.

Megan was terrified. If the man wasn't dead, she was in danger as well as Luke and her mother. It was the red which caught her eye as she turned to take refuge in the cabin. The blood had been covered with snow. Why? To make her doubt herself? The other body? Where was it? Megan found herself drawn to the lean-to. She followed a strange trail through the snow. There were no foot prints,

Steel Illusions

but a strange discrepancy in the pristine blanket of white. Megan held her breath as she look around the wood pile. She'd been hoping the body was there, hidden on the far side of the wood where she might not have seen it. But there was nothing. Megan backed away. Her light, what little there'd been, was diminishing, but her curiosity wouldn't let her give in. Eyes, trying to focus on the white stretched out before her, found nothing in the vicinity of the lean-to.

Megan walked out creating a larger circle of inspection. It was between two large Blue Spruce where she saw the imperfection. She deduced that the body in the lean-to had been moved to this spot; and then whoever had done the deed returned to the wood pile to cover his tracks with a pine bough. The trail she saw had one pronounced step of footprints; one step on each side of something dragged through the snow. *A body. The guy in the lean-to. No bodies - no evidence.* "Son-of-a-bitch, there's not going to be anyway to tie this kidnapping to Tulsa."

Fear had been replaced with frustration as Megan returned to the cabin. If this killer had enough strength to drag away a dead body, then he had enough strength to come after her. She had enough wits about to her to grab one more load of wood before taking the last few steps to the cabin door.

Megan practically jumped out of her skin when she had a clanging sound inside the cabin. "Who's there?"

"Meg, I was wondering when you'd get back. You must be freezing."

Megan was skeptical as she moved across the threshold and saw her mother standing in front of the shelf filled with canned goods. "Mother, are you okay?"

"I'm starving." Laura smiled as she met her daughter in the middle of the small room. She hugged the young woman tightly before asking if she'd like something to eat.

"Really, how are you?"

"My head's a little foggy, and my wrists and ankles hurt like hell; but it feels good to be moving around. How'd you find me?"

"Mom, I'll tell you while I fix us something to eat."

"Once a mother always a mother. You sit. I need to move. Is that Luke on the floor?"

Megan shook her head affirmatively. "Do you know what I'd love to be ordering right now."

Laura thought for a moment and was sure of her answer. "I agree. Tommyknockers. I want a Tundrabeary Ale and a Brewben sandwich."

Megan enjoyed playing the game with her mother. "I want the Maple Nut Brown and a big juicy burger. Man, a grilled burger sounds so good."

"What's that noise?" Laura heard a light taping sound near the window.

"I'm sure it's the Tommyknockers. You know the elves. They've come to protect us from danger." Megan watched her mother relax. Her logical mind deduced the noise to be a tree branch rubbing against the cabin wall; but she fantasized about the two foot tall, elf-like creatures who were supposedly brought to Colorado by Cornish miners during the late 1800's. The old stories credited the little creatures with saving many miners from disaster. "Mom, what do you suppose the Tommyknockers do now that there isn't much mining going on?"

"I guess they use their magic to save people like us."

"And make beer. Great beer."

"We could sure use a little of their magic right now. And the company would be fun too. With Luke passed out, a little male entertainment would help pass the time."

Steel Illusions

"Oh, Mother. That was a terrible pun." Laura looked at her daughter with a raised eyebrow. "A 'little' male company? You better be careful. If you insult them, they won't help us."

Laura gazed around her before saying, "Sorry, little guys. No pun intended." Laura heated the contents of two cans in a sauce pan while mother and daughter exchanged the details of their stories. It was the crackle of Luke's radio which interrupted them. "Hey, you guys out there anywhere?"

Megan moved to search Luke's body for the radio. She blushed as her hands moved across his body. The flush of her cheeks wasn't missed by her mother. Megan pushed the small button on the black box and answered, "We're here. Hey, Stew, is that you?"

"Yeah, what's going on. We thought you were going to radio us as soon as you got to the cabin."

"Things happened pretty fast."

"Why didn't Luke answer?"

"He's drunk." Megan snickered a little for effect. She was sure Luke would be just a little miffed at her remark. Serves him right for passing out.

"Drunk? What the hell are you two doing?"

Megan filled in the details for him and let Krista talk to Laura. Even without seeing her sister, Megan could picture the tears of relief rolling down her sister's face. It was in her voice as she asked if Laura'd been hurt. When Laura handed the radio back to Megan, her own eyes were beaming with tears. Megan reached out to squeeze her mother's hand before asking her next question.

"Did Jones make the drop for you?"

"Yeah, we're comfortable. Had a good meal considering the accommodations."

"How's Krista and the others?"

"Krista is doing well. Chelli's improving, but Brad is still in and out of consciousness."

"Why's he so bad? He seemed fine when we dug him out of the snow."

"Doc thinks it's a combination of being buried so long and then building up a sweat while he worked. His body temperature was already low, then sitting in his wet clothing lowered his temp even more. He's got all the symptoms of severe hypothermia. We're trying to keep him warm. The chopper should be in at first light. He'll be in a hospital by noon."

"Will somebody come and get us, or do we have to be at your location in the morning?"

"Stay put. We'll radio you just before we arrive. I'll radio Vail 1 that your mother is fine."

Megan looked at her mother. They hadn't talked about Laura's real pain. Megan had bandaged her mothers wrists and ankles while they traded war stories, but the was no mention of the intimate disgrace they both knew Laura had suffered. "Yeah, be sure to tell my Father not to fly out here. We'll meet him at home."

"Why?" Krista questioned in the background.

"Mom heard the kidnappers talking about him. They wanted him away from the mill."

"Are you sure you're okay out there. I don't like knowing the kidnappers could still cause trouble."

"One is dead, and I'm sure the other is badly hurt."

"What about Luke?"

"I don't know which will give him more discomfort in the morning, his arm or his head."

"When he wakes up, stay out of his way. I've only seen him with a hangover once, and he was mean enough to eat a bear."

"Warning taken. Sleep tight." Megan put the radio on the table and looked up at her mother.

"I know you understand what happened to me, Meg. Does he know?" Laura tilted her head toward Luke.

"Yes."

"I don't want anyone else to find out."

"Why? You should be checked by a doctor and tell the police."

"It wouldn't do any good. The man is gone; and even if the police find him, he'll be charged with murder and kidnapping. Rape doesn't need to be added to the list. It'd do me more harm than it'd do him."

"Even if the guy is found, I doubt if he'll be tried for murder. We have no evidence. He took the body with him. Probably buried it in the snow somewhere. "

"Then kidnapping will have to do." Laura didn't mention the other women Jake had told her about. She felt so dirty. *Control yourself. He's the guilty one.*

"You don't want Daddy to find out."

"That's right. It's over. I will not let the man destroy all of my life. What he did was bad enough, and I want to leave the disgrace here. Taking it out of the valley will ensure that it follows me the rest of my life."

"Mother, it will follow you anyway. You can't forget it."

"I can deal with it a lot easier if its not thrown at me over and over again. I don't want to live the rest of my life wondering what people are saying about me. Wondering if your father thinks about this man every time he touches me. I don't want to live like that. Can you even begin to understand?"

"Yes. I'd probably feel the same way. I wouldn't want...." she almost said "Luke" as she looked at him snoring peacefully on the floor. "....my husband knowing either."

"You two have made up I see."
"Yes, he even asked me to marry him."
"Really?"
"He was drunk. I don't think it counts."
"Me either." Laura smiled at her daughter. "Do you think he'll understand my request?"
"He's a good man, and he certainly appreciates personal tragedy. I think he'll respect your privacy to make your own decision."
"Do we dare go to sleep?"
"I'll put a few more logs on the fire. You try to get some rest, Mom. When Luke wakes up, I'll crawl in beside you."
"How can you stay awake?"
"There's some instant coffee on the shelf. That'll do."

****

Megan's head was resting on the table when Luke finally opened his eyes. He was sure she'd fallen asleep. Luke moaned as he tried to sit up. The pain in his head matched the pain in his arm. Niagara Falls was crashing down on his temples. Megan had said there was a bottle of aspirin in the medical kit, but the mission to retrieve them seemed impossible as he tried to push himself off the floor.

With several moments of "touchy feely" on the shelf and table, Luke finally gave in and woke Megan. "What? Are you sick?"
"I think so. I want the aspirin, but I can't find the bottle. I'm sorry to wake you up."
"I wasn't supposed to be sleeping."
"Why not?"
"The bad guy got away."
"What? How in the hell did he even get up? Are you sure? Damn!"

"Oh, yeah. How could I miss a dead body almost at the front door? Well, actually, I did miss it. It took me three trips to the wood pile before I realized he was gone; and, get this, he'd taken the dead body in the shed too."

"Son-of-a-bitch! I don't believe it. That means he's still out there." Luke was sobering fast. His arm and head still throbbed, but at least he was thinking straight. "How's your mother?"

"She's better. Let me get you a cup of coffee. It's all I could find except for the booze."

"I think I've had enough of that. Coffee sounds good. Aspirin too, please." Luke sat down at the table while Megan prepared two cups of coffee, and then slid into the chair beside him.

"Luke, she doesn't want you to say anything about...about how we found her."

"We have to, don't we?"

"I think it's up to her. I understand why. If something like that happened to me, I really don't think I'd want...." She nearly used his name again. *It's hormones. Just my hormones. I'll be fine when I get back to school*

"....your husband to find out."

"I guess so. How would you deal with it?"

"I honestly don't know. I'd probably want to kill the guy."

"That's understandable, but could you get past it every time you looked at your wife."

"I hope so. I think so. What happened to your mother wasn't her fault. She didn't ask to be dragged up here, drugged and tied to a bed. Sex as part of love is love. Sex as part of violence is violence. The two don't cross over. Sex is an act based on some type of reason. The reason is more important than the act. Does that make any sense to you?"

"I guess. Maybe one day, she'll be able to tell Daddy, but for now it is her choice. Agree?"

"Agree. Now what are we going to do about the kidnapper?"

"Get some dogs up here and track the son-of-a-bitch down. The chopper will be in to get us tomorrow."

"We won't be able to get up here with dogs until the spring."

"I already thought of that. I put a pair of gloves and a hat in your pack. Maybe the dogs will be able to get a scent from them."

"How do you know if they belong to him and not the other guy?"

"I don't. Just guessing. Maybe Mom will have some clues for us to go on. I want to get this guy."

"So do I. As long as he's up here, he's trouble. Does that mean you'll be here to help with the dogs."

"Absolutely."

"Did you talk to Stewart?"

"Yup. Told him you'd been partying and were drunker than a skunk."

"I deserve it. I can't believe I drank so much of that damn stuff. I'll never look at a bottle of Jack Daniels again."

"I'm sure." Her voice was full of skepticism. "You were funny. Your words slurred together and I had to almost carry you to the fur bed I made you. I'm sure your ski patrol buddies would love to see a video tape."

"I'm going to have to be weary of you, Sweety. I see blackmail is not beyond you."

"What are you talking about?"

"I wasn't so drunk that I don't remember asking you to marry me; and as I recall, you did accept."

Megan nearly spit her coffee in his lap. "You were kidding, under the influence. I won't hold you to a proposal like that."

Steel Illusions

"I was drunk, but I was serious. As a matter of fact, I doubt I'd have had the courage to mention the word marriage if I hadn't been drunk."

"Luke this conversation isn't going well. You shouldn't need courage to propose to a woman, only love."

She was right. He was making a shambles out of a tenuous friendship. "Megan, I've never been good putting my feelings into words. Not when it comes to women. My God, I spent most of my childhood with my grandfather. My sisters think of me as a guardian and a way out of the valley. I'm not their brother or their friend."

Megan watched his eyes but said nothing. She didn't know what to say. If she condoned his crass proposal, he'd think he could get away with it all the time. Her mother had always told her "what you see is what you get. You can change a man's clothes, but not the man."

Luke was determined to try again. This might be his only time alone with the most unique and loving woman he'd ever met. "Megan, I don't know if I'd recognize love if it jumped up and bit me on the nose. I've only loved Gramps and my dad. From what I've heard that's a lot different from loving a woman. Help me. Tell me what you want me to say." His eyes never wavered. He'd opened his soul to her; what she did with it was up to her.

Megan let woman's intuition guide her. She'd not taken her eyes from his. He held tightly to her hand as if he was afraid she'd pull away from his touch and his life. "I know I love you. I don't know if we can love each other. I grew up in a family where we all expressed our feeling openly. Daddy still tells Mom how much he loves her, right in front of us. Krista and I were always comfortable with their feelings for each other and with their love for us. I won't marry a man who can't tell me how he feels."

R. Z. Crompton

"Teach me. I know I want you. You're the only woman I've ever wanted." He leaned in closer letting his desire guide him. She didn't pull away. Her warmth was a magnet, more intoxicating than the drinks he'd had earlier. The kiss was light at first before his pent-up passion nearly devoured her.

"Jesus, Luke, I don't think you need any teaching." Megan was breathless when he finally released her. "I'm glad my mother was asleep."

"Will you marry me?"

"Yes. Do you still want an HIV test?"

"No, the results wouldn't matter. Tomorrow, before the others leave."

"Tomorrow? You mean get married tomorrow?" Megan was surprised by the sense of urgency. "Why, so quickly?"

"I'm more afraid of your walking away from me and never returning than the man out there waiting to kill me."

"Luke, my father has joked around about having his girls elope, so he wouldn't have to pay for a big wedding. Krista and I, however, want the big wedding. And, speaking of my father, I can't get married without him."

"My head, the one on my shoulders....." Luke smiled at her quizzical look. "Stewart filled me in on your analysis of doubled brained men."

"Oh," was her sheepish reply.

"My head is telling me to be patient. After all, I've been waiting a long time to find the right woman. A little longer shouldn't make any difference. By the way, I think I agree with you because right now my lower brain isn't agreeing with the one on my shoulders."

"I'll make a deal with you. I'm going to finish school." Megan watched the color drain from Luke's face. "Luke, please try to

Steel Illusions

understand, getting my degree from A&M has been my life long dream."

"So far, it doesn't sound like much of a deal."

"Please keep thinking with your upper brain; I'll address the lower one shortly." Megan squeezed his hand tightly.

"Come back with me tomorrow. Marry me in Tulsa, so my father can be present. We can have a short, but important honeymoon on my way back to school. The deal is, however, that for my graduation present I want a big wedding and reception with the fairy tale dress and the band for all my friends and relatives to enjoy."

"How long do we have to be apart?"

"You aren't exactly destitute. You can fly down to see me whenever you want. If I take summer courses, I might be able to get done in two more years. When I'm not in school, I'll come out here."

Luke didn't answer right away. Megan was actually afraid he might turn her down. "Where would we have the wedding?"

"Do I have to decide right now?" Megan answered Luke by getting out of her chair and moving to his lap. They shared a deep passionate kiss. Megan marveled at how wonderful if felt not having to second guess a man's intentions. She could finally enjoy the sexual pleasures she'd always denied herself.

"Are you sure you won't get jealous of my being out here with all this 'scenery' while your studying?"

"The only scenery I'm jealous of is your view of the mountains. If the other scenery didn't sway you before, then it won't now. Besides, we're either true to each other or not, a wedding ring won't guarantee fidelity. And a word of advice to you; jealousy is not an endearing trait." Megan couldn't hide her deep yawn from Luke. She was exhausted.

"Get some rest. I'll bring in more wood and keep a watch."

"Are you sure? What about your arm?"

R. Z. Crompton

"I'll be fine. You need to sleep. Tomorrow'll be another long day."

Megan didn't wait for him to change his mind. Her eyes were heavy, and crawling in next to Laura sounded wonderful. Her mother moved only slightly as she slide under the furs.

Luke yielded to his memories as he looked around the cabin. When he was nine and ten, this place had been his sanctuary. His father was so busy working; his mother hated the wilderness. Only Gramps had the desire to show him how to enjoy the serenity of the mountains. What his mother saw as wild and dangerous, Luke perceived as delicate and beautiful, easily destroyed by man's carelessness.

The arm throbbed when Luke pulled his snowsuit up over his shoulder. Holding a flashlight in his sore arm, he opened the door, surprised to see the sky full of stars. He marveled at the quiet and realized it was because the wind had stopped. The moon's reflection lit the entire area, and Luke gladly placed the flashlight back on the shelf. It wouldn't give him as much light as the moon would, and he'd be able to carry more wood if he didn't have to manage a light as well.

Luke stepped slowly watching and listening for any movement. He didn't like knowing a killer was still out here. *I know I hit the son-of-a-bitch hard. I can't believe he walked away and with a dead body to boot. Hell, I thought he was dead. Tough old buzzard.*

The crisp outside air was invigorating and helped to clear his still foggy head. It was on the third trip when he stop to rest his arm that he noticed the foot prints at the window. At first, Luke remembered he'd been standing there, but it'd been snowing then. The snow and wind should've covered his tracks. These tracks were recent. Luke looked through the small window again wondering about the view. *Hum? Strange odor. I wonder what it is.* A clear line of vision to the table, but no view of the bunk. The man who'd tried to

Steel Illusions

kill him had been back, watching them. He'd most likely been there scrutinizing Megan and her mother. Luke felt violated when he realized the man had probably watched him kiss Megan. With an eerie sense of déjà vu, he visually followed the tracks to where they disappeared into the trees.

Back in the cabin, Luke unsheathed the knife he'd found on the dead man's body. He couldn't afford to be caught off guard. He also rigged a covering for the window. *That asshole isn't going to spy on us!* Having boiled a pan of water, Luke made himself an extra strong cup of coffee and sat himself in a direct line with the door. The handle wouldn't even twitch without his catching the movement.

Waiting at the mercy of a mad man seemed stupid. The more the minutes passed, the more stupid Luke felt. His fighting ability was certainly in question. *Hell, this guy could start the cabin on fire and still end up killing all of us.* "Fire?" Luke smelled the finger tips of his glove. The strange scent he'd noticed outside at the window was kerosine. The walls of the cabin were covered with it.

Luke reached for his radio. "Stewart? Come in, Stew." He allowed a few minutes before repeating the call. His imagination had kicked into gear as he pictured the killer stalking Stewart and the others. *They don't even know he's out there.*

"Stewart," Luke yelled into the radio.

Megan rolled over on the bunk. She'd been asleep for less than an hour but still had trouble clearing her mind. "What's wrong, Luke?"

"I'm sorry to wake you, but I think you and your mother should get up and ready to travel."

"Why?"

"Gut instinct. Stewart, are you there?" He yelled into the small speaker.

"Yeah, yeah. What? It's only three o'clock. Did you want to redeem yourself by confessing the reason for your drunken state? Or are you going to try to convince me why I shouldn't tell a single living person that you went on a rescue mission and got falling down drunk so a beautiful woman had to take care of you?"

"Shut-up, Stew." The tone of his voice was all business and his friend recognized it from past adventures.

This time a more serious voice came back over the radio. "What's wrong?"

"Call Vail 1. Tell him it's clear and calm here, and we gotta get out, now." Luke was glad to see Megan and Laura moving so quickly. They were nearly dressed.

"Reason?"

"The killer's still out there. He's been watching us during the evening from the window. I found his tracks when I went out to get wood. We can't hike out, he'd pick us off one at a time. We aren't safe in the cabin either."

"You're safer there than outside."

"I don't think so. The walls have been doused with kerosine."

"I'll let you know how soon we can get a chopper up there. Get ready and watch for smoke. Hey, I'd open the door if I were you. You don't want him trapping you inside. I'll get back to you A.S.A.P."

Megan immediately stepped to open the door. Luke motioned, stopping her just as she reached for the handle. "What?" She asked. Luke put his finger to his lips telling her to be quiet. She heard the light scratching at the door. "Tommyknockers?"

Luke gave her a strange look, but Laura jumped in before he could speak.

"I don't think so." Laura answered pointing to the bottom of the door. Dried sprigs poked through the narrow opening.

"If he starts this cabin on fire, it'll go up in minutes. We gotta get out of here, Luke."

Steel Illusions

"I know, but not that way. He'll be waiting with a rifle to pick us off like ducks on a huntin' pond."

"We can't get through the window." Megan's fear was held at bay by her faith in Luke. "You know another way, don't you?"

"Gramps didn't put in many windows, but he did put in a back door."

Laura and Megan looked around but didn't see anything. They watched as Luke walked over to the bunk and kicked away the supports. The bed fell away revealing a small escape hatch that must've gone out near the lean-to.

"Gramps knew travelers sometimes needed the use of the cabin. It's pretty common up here to share. He didn't like the idea of poachers using the place but he couldn't stop it. He wanted to have something over on them. Something that proved he was smarter. I figured it was the eccentric part of him. He said the Tommyknockers told him to build it this way."

Laura and Megan smiled at each other. "See, I told you, Mom."

"What?" Luke asked.

"We heard them earlier."

"You sound as crazy as he did, and I almost dropped my teeth when you said the word earlier. Now, I not only believe Gramps may have been right about the Tommyknockers, but I don't think he was so eccentric either."

Luke opened the small hatch. There was nothing obstructing their exit. "We'll wait as long as we can. I hope Stewart calls back before we have to leave. I don't want to be out there running in circles."

He no sooner finished his sentence than the radio static caused them all to jump. "Come in, Luke."

"Hey, Stew. This guy's about to light a match. What's up?"

R. Z. Crompton

"Choppers are on the way back from Denver. Jones has already rigged his plane for an ice landing and is on his way."

"Jesus, he can't land his plane on the lake. It's not big enough."

"He'll be there in twenty minutes. You gotta get to the lake and be ready to load as soon as he lands. Landing won't be so bad, but he wants to do it on the first try. Going around again will give the killer too much time to figure out what's going on. Take off will be the tricky part with a guy shootin' at you, so he's not going to have time for a second go. Get in fast. Luke, you in the front, and the ladies in the back, one on each side. He needs the tail light and the weight balanced. Now get moving. It'll take you that long to get there without being seen."

"We're outta here. What about you guys?"

"The chopper will get us before the guy can hike this far up the mountain. See you back at base. Out. Now get going and, Luke, good luck." The black box was silent.

Luke went first to make sure the way was clear. He looked around the corner of the lean-to and motioned for Megan to move into the trees. "Stay low and quiet he whispered." The helmets had been left in the cabin. They were of no use now: to cumbersome and heavy. They didn't need any extra weight on the plane, even a helmet.

Megan was followed by Laura and then Luke as they maneuvered themselves slowly through the trees closest to the cabin. Once they got closer to the lake, they'd be able to move faster. It was the scent of burning wood which told them the cremation had started: their cremation. It wouldn't take the killer long to realize they'd escaped. When they didn't come running out the door, he'd look elsewhere for them.

Luke stopped briefly to turn and say "good-bye." So many happy days, good memories were burning to ash. Now he could remember with love and not sorrow as the tears flooded his eyes.

Steel Illusions

Megan reached for his arm. "We'll rebuild it. With more windows and a back door." He smiled down at her and then led her to the designated pick-up point.

By the time they reached the edge of the lake, the sound of the plane was audible. That sound would certainly tell the killer where they were headed. They stood straight against trunk of a tree, out of the killer's line of view, squinting at the night sky for the small plane. It came into sight just past the tips of the trees on the far side of the lake. Landing gracefully, the plane taxied up toward the shore closest to the waiting passengers.

Luke was sure "jerk number one" was headed for the plane by now. "Stay low and run!" He grabbed each of them by an arm not giving either a chance to hesitate. The shots started as soon as they cleared the trees.

"Shit!" Megan screamed as a shot zinged past her ear.

The plane was still moving as Luke pulled the door open. He reached over, first heaving Laura into the cabin and then Megan. Just as the plane started to make it's turn, a bullet caught Luke in the right thigh. Megan saw the flash of pain register in his eyes before his grip on the plane weakened.

"Mother!" She screamed. "Help me. Grab an arm. Quick!"

Jones hadn't said a word until both women reached for Luke. "Christ, you gotta hurry-up."

"Do you want us to leave him?" Megan snapped. "Pull."

"No, just hurry the fuck up, or we'll all get shot."

Megan didn't like his choice of words, but he was right. "Mom, get behind me. We'll do a whitewater save." Megan positioned her hands evenly on the collar of Luke's snowsuit. Gripping as tightly as possible and bracing her feet on each side of the doorway, she gave the final count. "On three, Mom, pull. One....Two...Three."

Luke flew into the plane landing on top of Megan and Laura.

R. Z. Crompton

As Steve turned the plane for its take-off run, Megan helped Luke get into his position. She went back to the door to pull it closed and looked out to see the killer standing on the bank with his rifle at his side. She gave a sarcastic farewell salute and pulled the door closed.

CHAPTER SEVENTEEN

Kevin looked over the papers Dan handed to him. "You were right, Dan. The figures here don't match what's in the computer. I can tell without even pulling up the numbers."

"Boss, what's going on. Why's Waterman trying to set you up?"

"I never said he was trying to set me up."

"You didn't have to. Those there papers did. I been readin' scrap reports for years. And those paper's er sayin' you been stealin' from the scrap pile."

"How so?" Kevin was curious about Dan's astute assessment of the situation.

Dan was in the middle of his explanation when Rick entered the office. It was nice to just drive through the front gate this time. "Thanks for leaving me the clearance, Kevin. I'll have to try it more often."

"If you'd just call first, I'd leave you a pass."

"I thought it was my police escort. This is Walt Andrews." As Rick made the introduction, a business like handshake was exchanged. "He might be able to give us some help."

"Dan's giving me his perspective of this whole situation. I've got to give him credit; he's been a lot quicker than I have at understanding this mess."

"Hell, Boss. If you'd ask me first I'd a told ya a thing or three about this place."

"I'm sure you could. The story I want to hear now is about the workers most likely to steal from this company."

"Stealing from a steel mill. Sounds kinda funny doesn't it, Boss."

"Yeah, it creates quite an illusion. They've got their own kind of steal company running within a steel company."

Steel Illusions

"I'm not so good with words, but I know I don't like the picture you're making. Good steel men, the guys I know and work with, don't steal from each other or the company."

"I know, Dan. That's why it took Rick so long to convince me. So tell me what you think. Waterman didn't pull this off alone. He's got guys on the inside."

"Yeah, he does. Sam gets my first vote. He checks the scrap in and would have plenty of time to fix the records."

"And lie about what was in the load?"

"Sure, why not? Pete wasn't part of this. He was used. Sam could have easily sent in a bad load. Pete see's something wrong, runs down to check it out, and bang."

"I don't know how he fits into this mess, but that's my feeling too." Kevin agreed. "The visitation for Pete is tomorrow. Are you going?"

"Yeah, me, Betty, and several others."

"Good. I can't go. I'm afraid I'll upset Amy too much. Please give my regards."

"I'll talk to Amy myself."

"Good luck." Rick interjected. "I've been trying to see her all day. Her mother's like a warden."

"My Betty is a friend of hers. I'll work on Amy, and Betty'll check out her mother. If they know or even suspect anything, we'll find out. There's nothing Betty can't weasel out of somebody."

"Rick, he's right about that. The woman's a legend in her own time."

Kevin and Rick pieced together the limited parts of the puzzle for Walt and asked his opinion. "Keep your head down, guys. Snooping around can cause big trouble. You don't have enough yet for us to charge Waterman with anything. If he's as cocky as you say, he'll make a mistake."

"Thanks for coming out, Walt. We needed some advice."

"Any time you have a question or need back-up just call. The explosion isn't under investigation by the police. Should it be?"

"Yes." Dan answered before Kevin could rationalize giving an affirmative nod. Rick was surprised Dan would reply at all. It really wasn't his responsibility to speak for the company.

"Why?" Kevin and Rick stared at Walt and then Dan, waiting for the response.

"Because a bunch of us think Pete was murdered."

"Jesus, Dan. You can't just come out and say something like that."

"Why not?" Walt asked. "How do you think we find out about most crimes?"

"Why do you think he was murdered, Dan." Rick's reporter instinct kicked in.

"Cause the kid wasn't stupid or careless. I trained him myself. He was down there because he saw or knew something was wrong. We just haven't figured out what it was yet. The scrap was bad; it caused an explosion, Sam was the one who checked it in. Now I've got papers, which I know were planted. These papers make it look like my boss is stealing from the mill. Not a chance! I saw the man who fixed these reports."

"Guys, I agree with Dan. It's time to start asking questions."

"What if it scares Waterman into pulling back?"

"He sounds pretty pompous. He probably thinks he's to smart for us. What about the wallet you two found in the scrap pile? Any luck finding an owner?"

"Not a thing. I passed the name around but got nothing. If he was hired from the other side, the scrap side of this mess, it's possible nobody around here knew him."

"Or they're very good actors." Rick added.

"I doubt it. Good liar, maybe. But nobody around here can act." Dan smiled at the image of a steelworker acting on a stage. *Not a chance!*

"So what's next?" Kevin asked.

"Let's make a copy of these pages and print me out what is really in the computer. Let's play the game out a few more moves. Put the papers back where you got them. I'll keep a copy. Shit, make everybody a copy. When Waterman plays his hand and pulls out the incriminating evidence, you can wave your own copy in his face. Dan you're the witness who saw him in the office. Don't tell anybody. He'll want to silence you. At this point, he might try anything."

"I'd feel better if you and Betty stayed with your brother for a few days. I don't think it's a good idea for Betty to be home alone. Take the weekend off."

"I can't do that. There's work to be done out here. Who's going to take care of this place."

"Shit, Dan. We're only on one furnace. The rest of the clean-up can wait 'til we get this mess figured out. Now take Saturday and Sunday off. That's not a request."

"Rick, see if you can hurry-up your friend in the east with that report. When Waterman decides to accuse me of something, he'll try to make a big deal of it: press and lots of people. If I keep a low profile for a day or two maybe we can have that report and something from Amy."

Rick was impressed with Kevin's orders. He was finally taking control of the situation by adding some of his own rules to the game. "Kevin, where will you be?"

"Waterman can't get me in front of the press if I'm not in town. I'm going to take the first flight I can get into Colorado Springs and collect my family. I hope. I'll be back late tomorrow night."

"What about the danger out there to Laura and the girls?"

"At this point, there's as much danger here as there. The girls will be flown in tomorrow. I'm going to be there. Waterman's game is printed out in several copies. When we pull the plug on him, I want Laura standing beside me. I'll call you as soon as I get back. We'll decide our next move."

****

Steve Jones glided in for a landing at his private landing strip. An ambulance was waiting to take the passengers to the Vail Valley Medical Center. Luke was conscious but suffering from serious loss of blood. The bullet had hit an artery before exiting through the front of his leg. After being loaded onto the stretcher, Luke shook the older man's hand.

"Thanks for coming to get us, Steve."

"Yeah, well you owe me like hell, Kid."

Luke smiled, "I know; I seem to fall further into debt with you every time I see you."

"Hell, you could buy me a new plane by now." The man smiled at Luke.

"Well, I'd rather owe you than cheat you."

"Get him outta here." Steve yelled at the driver.

Luke was loaded into the ambulance while Megan offered her own thanks. "My mother and I thank you too, Sir."

"Shit, it was nothing. Just a little entertainment for a rather boring Friday."

Mr. Jones was rough on the outside, but Megan read him clearly. She reached up and kissed him on the cheek. The ruby red above the unshaved cheeks showed his deeper tenderness. "Go on, get outta here." He watched the white and red vehicle until the lights were no longer visible and then touched the side of his face.

Steel Illusions

"Megan, when we get to the hospital..." Laura wasn't sure how to word her request. She didn't know if Luke was willing to go along with her secret or not, and she had only minutes to make sure they all told the same story."

Megan anticipated her mother's request. The press and police would be waiting for interviews. They needed to be thinking alike. "Mother, you can use whatever facts you want. Luke and I found you drugged and tied up. That's all. We never really saw anything. Isn't that right, Luke?"

"It's your story, Mrs. Bradford."

"Please, call me Laura, or Mom."

"You heard us?" Megan blushed.

"Sure did. How could I not? It's going to be hard for Luke to make the trip to Tulsa and be in the hospital at the same time."

"I hadn't thought of that." Megan said as she looked down at Luke. "What are we going to do."

"Treat this like any other mission: wait and see what the options are."

"I never thought of getting married as a mission. It sounds so romantic."

"In this case, it is a mission. I'll make up the romance part after I know you're mine."

The ambulance pulled into the clinic scattering a throng of reporters. The driver hadn't expected people to be standing in the middle of the driveway. His sudden swerving nearly threw Luke off the stretcher. Two technicians jerked open the doors and pulled Luke out of the vehicle.

Tim was waiting for all of them as they were ushered through the doors. Luke talked as fast as the technicians walked trying to give Tim the pertinent information. "Go get yourself fixed up. You're gettin' blood all over." Tim slapped Luke on the shoulder.

Luke winced in pain. "Ouch!"

"I thought you were only shot in the leg."

Megan, following closely behind the stretcher, offered the quick explanation. "The jerk got him with a knife first."

"Sorry, Miss. Only family beyond this point."

Megan started to protest, but Luke interrupted, "She is my family."

"Sorry, Sir. We have to follow the rules."

"Follow this!" Luke snapped. His temper was flaring according to the pain. "I paid for most of this building; and if you want a job tomorrow, you'd best consider this woman my wife. Do you follow?" The words were short and hard.

"Yes, Sir, Mr. Halverson. I follow." The poor intimidated man whisked Luke away with Megan at his side.

"Wife?" Tim looked at Laura with the question repeating over in his mind as he turned to look again at the man being taken down the hallway. "What happened up there? I was beginning to think nobody would crack his shell."

"She not only cracked it; she scrambled him up pretty good. Don't I know you from somewhere?"

"Yes, Mrs. Bradford. I've met you when you stayed at Montaneros."

"Of course, I'm sorry I didn't recognize you."

"We look different in our different roles. I'd much rather be seeing you under the other circumstances."

"Me too. I suppose you want to ask some questions. Is there a better place to talk?" Laura liked the Medical Center. It wasn't as overpowering as the big trauma centers in the city. She was hoping the treatment practices were more personable also. "The reporters will only be held at bay for so long."

Steel Illusions

Tim led her to a small room with large comfortable chairs and a sofa. "Would you like a cup of coffee?"

"Actually, I'd rather have something stronger. It's been a hell of a day."

"How 'bout coffee now. Hospitals aren't known for having well stocked bars."

She smiled at the friendly face. She hadn't really expected a good bottle of brandy to be stuck in his pocket. "Black, please."

Within minutes Tim was walking toward her with a cup of steaming brew in each hand. "Hope you like it strong. I think it's been sitting there all evening. Do you need medical treatment? Are you hurt?"

"My wrist and ankles are pretty bad. I thought I had some frostbite; but after running to the plane, I guess I'm okay."

"May I see?"

Laura was glad to get the snowsuit off. She'd been in the same clothes for nearly forty-eight hours. A shower would feel wonderful. Pulling up the arms of her sweater and removing the bandages Megan had used was easy. Tim looked tenderly at the deep raw marks on her wrists. Then he looked back at her.

"You have these marks on your ankles too?"

"Yes. Why?"

"You were tied to....?"

"A chair." There was no hesitation in Laura's answer. She'd decided before getting off the plane how she'd answer the question.

Tim looked her in the eye waiting for a different answer. The torn flesh was not from being tied to a chair. He'd seen plenty of rope burns before. Doing rescue work, he'd suffered with his own burns. It was never this bad. She'd have scares there for the rest of her life. There was no skin to even stitch back together.

"You need a doctor alright." He hesitated before adding, "Are you sure you don't want to see a gynecologist?"

"No." The answer was flat and final.

"This didn't happen by being tied to a chair. You were tied to the bed."

"It doesn't matter what you think or guess. As far as everyone is concerned, I was tied to a chair. The ropes were too tight and this is what resulted." Her eyes were fixed on his. Her tone didn't waver. There were no tears of shame only the resolution that her facts, as told, would be the facts of record. "Anything else would serve no purpose. Can you possibly understand?"

"Yeah. May I suggest you go to your personal doctor when you get home?"

"Don't worry. I plan to. I can live with the bad memories but not with anything else he might have given me. You won't tell what you suspect?"

Tim shook his head. "I won't tell what I know either. Can you give me a description or name?"

Laura told him as much as she could about the abduction, the trip through the wilderness and the two men. "I recognized one of the guys as the man who was at Cassidy's."

"Had you ever seen him before?"

"I don't think so. You know how you think you might recognize a person, but you're not sure? Well I'm not sure. Shave the beard and put on different clothes, he could look very different. We don't even have a body to look at. I only saw him once in the cabin and I was face down on the floor trying to get a look at the guy in the dark. Not much to count on. The other creep, Jake. Shit, if you ever find him, he deserves to be turned into 'Rockie Mountain Oysters': chopped up into tiny pieces and deep-fried."

Steel Illusions

Tim couldn't hold his laughter. "I'm sorry, I know this isn't funny, but there's more to our Rockie Mountain Oysters than just deep frying little pieces."

"I know what the little pieces are, Tim; and believe me, that's one body part I'd cut up with pleasure."

"Would you recognize him?"

"Before or after I cut him up?"

Again he couldn't hide from her sarcasm and smiled boldly. "Before."

"Most definitely. The man was planning to kill me; he didn't bother hiding his face. With a police artist, I could give you a pretty good picture."

"Good. Now let's find you a doctor."

A pretty young woman in a traditional doctor's coat came into the lounge where they'd been talking. "Mrs. Bradford, your daughter tells me your wrists and ankles need some attention."

"She's all yours, Doctor. We're done here." Tim got up to leave.

"Thanks, Tim. For everything."

"No problem. I understand. I'll probably have more questions for you later."

"I'll be here. When will Krista and the others be back?"

"Any minute. We'll bring them straight to the Med. Center."

Laura was led to a small examination room where she undressed and pulled on the always elegant, white hospital gown with the little ties in the back and the good ventilation system. She was sitting on the cold, hard exam table when the door opened. The man walking in surprised her. She'd been expecting the same doctor who'd escorted her to the room. She looked ambiguously at the man walking toward her. He didn't look right. *Something isn't quite right?*

"Mrs. Bradford, can you explain your injuries to me?"

R. Z. Crompton

Strange, the other doctor already knew what her injuries were. Oh, well. There's no reason not to show him. Laura held out her arms for the initial inspection.

"Wow! How'd that happen? Must hurt like hell. "

Well that certainly wasn't the most professional response. "Excuse, me do you have some I.D?"

Patting his coat pockets, the man said, "Guess not. Must'a forgot it in my other coat."

Then Laura saw the notebook in the pocket. It was scruffy looking with dog-eared corners. "You're a reporter, aren't you?"

"How'd you guess?"

"The notebook. Next time you sneak into a place, leave it in the car. How'd you get past the others?

"Easy. White coat. They didn't notice the pad of paper." He reached down to touch his security blanket. "Can't leave home without it."

"What's your name?"

"Dan. Just Dan. Will you answer a couple of questions for me? Please."

Laura eyed him carefully. She'd have to talk to the press eventually. Might as well be the reporter with the most initiative. "Which paper?"

"The local one. Does that matter?"

"Yes. It means you know the area. I'll make you a deal. I've got two daughters and the other members of the rescue team which risked their own lives to save me from being murdered. We're craving a special treat. If you'll drive into Idaho Springs to Tommyknockers and bring back enough beer and sandwiches for seven people, eight if you want to eat with us, I'll give you the whole story, and you can talk to the rescue team. I know it's a long drive, but there's no way I can get out of here for a while."

Steel Illusions

"I don't know. How do I know you're not going to talk to somebody else?"

Before Laura could finish her transaction, the doctor she'd been waiting for opened the door. "Who are you?"

Laura made the introduction. "This is Dan. He's going to fill an order for me." She looked back at the young man, intent on getting her way. "And to answer your question, you don't know if I'll talk, but then you'll have my food. And I know I don't have the energy to tell the story until I eat."

"It'll take me two hours to get over there and back."

The young lady's head moved from Laura to Dan before interrupting. "What food? We can get you something to eat here."

Laura stayed focused on the man in front of her. "I know how long it will take. I'll call the manager, place the order and pay for it. If you want the story, you'll come back with my food."

"Is this some kind of blackmail?"

"I'd rather think of it as the ultimate in the barter system. You want something; I want something. It's a mutual exchange."

"I don't think so, Lady. I get the two hour drive."

"Let me check outside. There must be somebody who wants this story."

Dan smirked, "Now that's blackmail. I'll do it. Where will I find you."

"We'll still be here, I'm sure. Two members of the rescue team are hurt and haven't arrived yet."

"Then a full story?"

"As much as you can write."

Dan hurried past the doctor and out of the room. "Why are you going to talk to him?" She asked Laura.

"Why not? I've got to talk to one of them eventually. I'll tell Dan the story, and all the others can copy his. How long will it take to bandage these? I have an order to place." Laura held her wrist and ankles out to the doctor standing in front of her.

"These are nasty. We'll get you cleaned up, but you'll need care from a plastic surgeon when you get home. I'll do your ankles first. Here's my cellular."

"Thank you. What about the young man who was brought in?"

"He's being stitched up."

"Mother!" Krista ran into the room and threw her arms around the woman still sitting on the table. They shared tears of relief and details of their ordeal while the doctor finished her job.

"When do you think we'll be able to get out of here?" Krista asked her mother.

"Sometime tomorrow, I hope. I want to make sure Brad's okay. Have you seen Megan?"

"Not yet. Why? What's the strange grin for?"

"I can't tell you. She get's to tell her own news."

"What? Did she and Luke make up? I hope so. She was a pain in the ass while they were fighting."

"Krista! Don't talk that way."

"Really, Mom. She had a major 'itch' attack."

When Laura was bandaged, she got off the table and pulled her dirty clothes back on. "Man, I want a warm shower and clean clothes."

"Ah, I'm sorry, Ma'am." The doctor replied. "The clean clothes are fine, but you can't get the bandages wet for a couple of days."

"You're joking, right? A cruel joke."

"Wait until you see a plastic surgeon, or you might have terrible scars. It's not so bad if you're living in ski clothes, but you won't like it in a swimsuit."

"You've made your point, thanks, Doctor. C'mon, let's check on Chel and Brad." Laura and Krista headed out the door nearly colliding with the man racing down the hallway.

Steel Illusions

"Kevin? Kevin! How?" Her arms were around his neck and her feet off the floor. Kevin swirled Laura through the air as she kissed his face and neck.

"My God! I didn't think I'd ever see you again." Kevin didn't see his daughter immediately. It was only after swinging Laura through the air that he saw her standing beside them. "Krista, you're here too. You were in an avalanche."

"I'm fine, Daddy." Her own tears mingled with his as she kissed his check and soaked up the bear hug.

"Meg, where's my baby?"

"Let's go find her." Laura suggested. Kevin wrapped an arm around the waist of each woman nearly carrying them down the hallway.

"We'll surprise her." Krista peeked into each room looking for her sister. It took only three tries before she found Megan sitting next to a very handsome man in a hospital bed.

Kevin went in first. "Miss, you'll have to leave."

Megan recognized the voice just before she turned around to object. "Daddy? Daddy, I can't believe you got here. We thought you couldn't leave."

"It's a good thing I did. Obviously, you're still in danger." Kevin said as he looked at the man in the bed.

"Oh, Dad. Stop it! This is Luke."

Luke tried to sit up, but he couldn't shake the drowsiness of the sleeping pill. "Hello, Sir." A weak hand came up off the bed in an attempt to offer a handshake.

"We'll talk tomorrow, young man. Then I'll want to know your intentions."

"Dad! This is no time to joke around."

"Who's joking. Anytime a woman hover's over a man in bed, it's time to ask what his intentions are."

"Marriage, Sir."

Krista was the first to react. "Meg, I guess you guys really made up in the woods."

"Krista, don't put any ideas in his head." Megan motioned to her father. "He's bad enough without your help."

"Meg, what's going on?" Kevin looked at his youngest daughter with only love in his eyes. He'd just found her only to lose her to another man.

"Oh, Daddy. I love you." She ran and threw her arm around him. "I know you're just teasing me. He's a wonderful man."

"I'm sure he is, or you wouldn't give him the time of day. When's the wedding?"

"Tomorrow."

"Tomorrow?" This time the reaction from Kevin, Laura and Krista was in unison.

"He won't even be able to get out of bed tomorrow." Krista argued.

"What about school?" Kevin threw in.

"No big wedding ceremony?" Laura's disappointment was evident in her voice.

"Do you want the answers, or are you going to keep yelling at me? We were going to go to Tulsa just so Dad could be with us, but that's going to be impossible now. Daddy, you're a saint. You solved the first problem. Thank you for coming."

"I guess I should've waited."

"Don't be silly. Since I'm sure you want to get home as soon as possible, Mom, Luke and I will use the car and drive me back to school. I'll finish school at A&M. We'll travel back and forth when we can."

"You've been busy making lots of plans."

"C'mon, Mom. What were you going to do with the car? Leave it here?"

"I hadn't gotten past wanting something to eat."

"As for the wedding. It'll be my graduation present from Luke. In Tulsa. Just to answer your next question, Mother."

"Don't forget a reception in Vail too." A weak voice filtered into the family conversation from the bed.

"Your supper has arrived, Mrs. Bradford." Dan walked into the room with arms full of white bags.

"Dan, come in. We're having a rehearsal dinner."

"What? Beer and sandwiches?"

"Yes, but they're Tommyknockers. We better find Chel and Stewart. They are part of this party too."

"What about Brad?"

"He won't be partying for a few days, but at least he's going home tomorrow."

"Excuse me, Mrs. Bradford. You promised me a story."

"Yes I did, and you'll get a big one. Have a seat and a sandwich."

CHAPTER EIGHTEEN

Kevin unlocked the utility room door and punched in the code to turn off the alarm. Because of the rush he'd been in to catch the last plane into Colorado Springs, he didn't have any luggage; but Laura and Krista certainly made up for his missing share. He'd been surprised the airline even checked all of it.

Megan and Luke were going to use the last night at the condo and leave for Texas early Sunday morning. Kevin still didn't like the reality of his baby being married. His head understood she was twenty-three, but in his heart, she was still his baby. He was jealous of this tall, handsome millionaire taking his daughter away from him.

Dan had rushed to get his story to press after the rehearsal dinner and then made sure Kevin and Laura had a copy before leaving late Saturday night. He had written a nice story about the small wedding, announcing the catch of the valley's most desirable bachelor.

Laura was already in the bedroom unpacking her large suitcase when Kevin came through the door to check the phone messages.

"Honey, some guy named Rick called. It was a strange message. Dan called too. Man, it feels good to be home."

Kevin walked over to hug her again. "I wish Meg had come with us."

"Stop pouting. You knew it'd happen sooner or later."

"Later is better. At least he's well-off. He'll be able to afford to keep good beer in the refrigerator and fine wine in the cellar when we visit."

"Please understand if I don't go out there in the near future."

"Don't blame you one bit, My Dear." Kevin answered as he pushed down the button on the answering machine.

Steel Illusions

The machine responded with static and then a soft whisper like sound identified the caller as Rick. "Kevin, I....I got ne..." there was a long pause before the word was spoken, "news. I'm in...your...."

"What? In my what?"

The next message was clear and quick. "Boss, call me."

Kevin picked up the phone to dial as Krista walked into the room, "Dad, where's the dog?"

"At the vet, Honey. I didn't know when I'd be coming and going from the house, so I took him to the clinic."

"I'll go over to get him tomorrow. Anybody want anything to eat?"

Kevin had forgotten about groceries; and as he dialed Dan's number, offered, "Sorry, you better order a pizza. I don't think there's much...Dan! We just walked in the door. What's up."

"Talked to Amy today. She knows somethin' but ain't talkin'."

"How about her mother?"

"It's the damndest thing, she ain't talkin' either. Betty's sure they're hidin' in shit up to their eyeballs."

Kevin smiled in spite of the seriousness of Dan's information. The man's colorful vocabulary constantly entertained him. "Did you talk to Rick? He left a weird message here."

"AHHHHHHHHH!!"

Kevin dropped the phone and ran for the kitchen right behind Laura. They saw Krista frozen in the middle of the room staring out the back window. Laura reached her first and followed her gaze.

"AHHHHHHHHH!!" was the like response.

Kevin went directly to the patio door and peered through the glass. "On, my God." Unlocking the door, he pulled it open and rushed out. "Rick? Rick?"

"Kev, is it you, Buddy?"

Laura had regained her wits and was standing behind Kevin as he knelt down beside the man slouched in the patio chair. "Call 911! Hurry!"

Laura went for the phone and found Dan still on the other end. "Rick's hurt, Dan. Gotta go." Laura hung up the phone but had to go back into the bedroom where the other phone was still off the hook to clear the line and get a dial tone.

"Got the news..." he coughed. Kevin looked over his body trying to find the cause of his injury but found nothing.

"...from my friend."

"Don't talk; Laura's calling the ambulance."

"No. Call Walt. It's Wat..." he ran out of breath before he could finish. "Waterman for sure. He shot me."

"Where are you hurt?"

"Cylinders..." This time Kevin waited for Rick to take a shallow breath and continue. "...were bought in Sand Springs two weeks ago."

"Two weeks ago?" Kevin was stunned. He still hadn't really believed anybody would intentionally blow up the meltshop. This was the proof he needed. "Where's the report? Rick. Rick?"

There was no answer, only a siren could be heard screaming down the street. Laura stepped back to the door. "Is he dead?"

"Not yet. Show the paramedics in."

The men worked quickly to get Rick ready for transport. Kevin hadn't seen any blood and even when the body was removed from the chair there wasn't much blood. He'd been shot in the back at close range.

"Sir, did you call the police?" The medic asked.

"No we just got home and found him here."

"I gotta radio this in. This is a crime scene. I don't know if I can take him before the police get here."

"What in the hell are you talking about?" Kevin yelled. "You'd let the man die just because the police aren't here? That's ridiculous."

"That's orders. We call in. They tell us what to do."

The man made his call and the police were dispatched to the scene. Kevin felt less agitated when he heard a doctor command "transport immediately."

Guilt beat on Kevin's conscience. Rick had stepped in as a friend to help him, and now he was dying. Waterman would pay for this, but first Kevin needed to find that report.

Kevin watched Rick being loaded into the ambulance and replayed the brief conversation. *Call Walt.* That was one instruction he could follow easy enough; but before he could find the number, another line of the conversation played over in his mind. *Waterman did it. He could be here.* "Here? Shit!" *He must've met Rick outside. The house was still locked and the alarm set. But he must'a been here and with a gun.* "Laura, Krista where are you?" Kevin nearly knocked Krista over as he rushed down the hallway.

"What, Dad? I didn't want to be in the way, so I went up stairs. What's wrong?"

"Stay away from the windows." He ordered while he checked and rechecked the three doors which allowed access to the house.

"Kevin, what's wrong?" Laura demanded.

Kevin was frantically looking through his wallet for the card Walt had given each of them the night before. "Waterman was here; he shot Rick."

"Aren't the police on the way?"

"Yeah, but they'll think I did it unless I can get ahold of Walt. And I won't get a chance to call him once the police get here and start the inquisition."

"This guy has weaved one nightmare after another for us, hasn't he? Did he do the kidnapping too?"

"I'm sure he did." The card was the last one he pulled out. Kevin reached for the phone and heard the doorbell ring at the same time. "You answer the door. Try to stall just for a few minutes. Do the introduction thing slowly."

Laura walked nervously to the door trying to count seconds between her steps. "Thank you for coming so quickly...." She said when she opened the door.

Kevin heard her go into a drawn out explanation about their arrival and discovery of the body on the patio. The phone rang, and rang again. He heard Krista now telling her part of the horrible story. The phone rang again. Kevin was mentally formulating the message he would leave and then panicked at the possibility that Walt might have been shot too. An officer was coming into the kitchen when Kevin finally heard the voice he'd been waiting for.

"Walt, Rick's been shot at my house by Waterman."

"Sounds a little bit like Professor Plum, with the candlestick, in the Billiard room. Don't you think?

"Sorry, I'm out of humor. Can you come over right away?"

"Is there an officer there now?"

"Yeah, just arrived."

"Let me speak at 'im for a second."

Kevin handed the phone to the lady coming toward him. She didn't look happy at all to find him talking on the phone. Laura's stall tactics had been a little to obvious. "It's for you."

After a couple of "Yes, Sirs" and "No, Sirs" she handed the phone back to Kevin. "Detective Andrews wants us to call homicide. He'll be here right away. Harry, you go around back. We gotta make sure the killer isn't still around."

The other man looked at her and then back at Kevin. "But what about him."

"I guess he's not a suspect." The officer coldly added, "It must be nice to have friends on the force."

"You have no idea."

There was no response as the officer turned to go back to her car. Kevin walked to the wet bar in the family room. The emotional bungy cord he'd been swinging from the last four days had him primed for a good, stiff drink.

"Me, too." Laura said walking up behind him. "Are we prisoners here like we were in the cabin? Waiting for the killer to come and get us?

"No, Dear." He reached out and pulled his trembling wife into his arms. "The police will help us. Waterman can't hide forever. The arrogant son-of-a-bitch probably doesn't even suspect Rick identified him. Here, drink this." He handed her the brandy snifter.

"Hey, guys, do you have one of those for me?" Walt walked into the room.

"I thought you couldn't drink if you're on duty."

"I'm not, officially."

After introducing Laura, Kevin expressed his concern. "I'm worried about Dan. He thinks he's invincible."

"I know. I've sent a car out to his place."

"Thank's. What do we do?"

"Sit tight. We'll take care of it. We're already looking for Waterman. He can't hide forever."

An officer came into the room and whispered something to Walt. He was instantly on edge and started issuing orders. Kevin didn't like the sound of sending squad cars to the melt shop. "Why?" Kevin tried to interrupted.

Walt didn't answer right away, but eventually addressed Kevin's concern. "We can't find Dan. He's not at home. Rick's alive but it doesn't look good. Where are you going?"

R. Z. Crompton

Kevin had gulped the rest of his brandy and started out of the room. "The mill. If Dan's gone to the shop, there's a million places he could be."

"I'll give you a ride." Walt followed Kevin out the front door.

"Hey?" Laura yelled after them. "What about us?"

"You'll be fine, Mrs. Bradford. A police officer will stay outside after the others are finished out back. It won't be much longer."

"Damn!" Laura's own frustration hit a new high. She poured herself another glass of brandy and shuffled off to her bedroom. Alone. She couldn't take a warm bath, so she might as well bathe her emotions in something warm. The brandy would help her sleep, and that's what she needed. Kevin might be gone all night.

****

Walt pulled into the lot right behind three other squad cars. Kevin ordered the guard to open the gate. He didn't need to ask if Dan or Waterman were there; both cars were in the parking lot.

Kevin searched through the shop with officers spreading out all around him. Nothing. Dan hadn't even been seen by any of the guys on duty. "Where in the hell could he be? The scrap piles! C'mon, Walt." Kevin ran to the car.

Directing Walt through the buildings to the back lot, Kevin's heart raced. If Waterman killed Dan, there'd be hell to pay. Betty'd barbecue his ass in his own furnace. "Hurry!" The car spun through the gravel throwing small chunks at the vehicle behind them.

"There!" Kevin pointed to a slight movement in the darkness. "Go to the right." The car raced to the right and stopped. Two figures were standing beside a large mountain of scrap.

"Stay back. I'll shoot him."

"Kevin, is that Waterman?"

Steel Illusions

"Yeah."

"He's got nothing to loose by pulling the trigger. Is there any way we can get around him?"

"He's backed up against the fence. You'd have to drive all the way around and come in from GCR's property. I'd have to go cause you'd never find it in the dark. Shit, I don't know if I can find it in the dark."

"Do you have another suggestion?"

"Can't you talk him out of it? I thought that's what the police were supposed to do."

"Do you think this guy is going to throw down that gun and walk away?"

"Shoot him straight out."

"You've seen too many movies."

"He can't kill Dan without losing his shield. Ask him what he wants."

"He wants to get away."

"So make him an offer. It might buy us some time."

"Waterman!" Walt yelled. "Let him go."

"I'll trade him for Bradford." Waterman screamed at them. "He's the one I want."

Walt was shaking his head when Kevin answered, "Okay."

"What? Are you kidding? That's the worst thing you could do. He'll kill you just for the pleasure of it. You were the center of all this in the first place. Don't play into his hands."

"I can't let him kill Dan. You were right. This is my fight; not Dan's or Rick's. Rick's dying. I won't let the same thing happen to Dan."

"I gotta an idea. How's Dan's hearing?"

"Terrible. If his hearing aids aren't turned up, he won't hear a thing."

"Shit."

"What's the plan?"

Walt grabbed a couple of officers and tried to explain what to do. "You game, Kevin?"

"Yeah, don't worry. I'll get Dan."

"Kevin, try to get him talking. His finger won't be so itchy if he's talking, he's not so likely to see one of my guys moving."

"Waterman!" Walt yelled into the darkness. "Bradford will start over, you let Dan go at the same time. Even exchange."

"If I suspect anything, he's dead. Tell those cops to back off."

Walt motioned for his men to move back. All of them within Waterman's line of vision moved away. It was the officers already behind Walt's car with the high powered night rifles who didn't move. Not a muscle twitched as they focused in on the tall man behind the human shield.

Kevin walked at a steady pace toward the man who wanted him dead. He'd been afraid Laura or one of the girls would be killed in the mountains; and, now, it looked like he was going to be the one to die. Waterman's gun was no longer being forced into Dan's back. Kevin could see it aimed at him.

"Why ya do it, Waterman?" *What an idiot question. He did it for the money.*"

"To get rid of you, you righteous son-of-a-bitch."

Now Kevin was curious. "Why? You hired me."

"You didn't play by my rules. You questioned everything."

"Then why in the hell didn't you just fire me?" Kevin's heart was pounding in his chest.

"You had to go out in disgrace. You were the hero; I needed to destroy you. You were supposed to leave for the mountains while all the details were put in place." Waterman moved his free hand to wipe the sweat from his brow.

"You used too many players. You can't control everybody."

"If I'm going to hell, I'm taking you down before I go." Waterman held the gun directly at Kevin's chest.

Kevin looked at Dan. His stride was rigid, but steady. He was almost to him. Three more steps....two.....one. Kevin dove to his side grabbing Dan at the hips and pulling him to the ground.

Shots were fired before the two men hit the ground. Only one bullet struck the ground near Kevin's shoulder. The others had been searching for a different target. Kevin lifted his head and saw Waterman face down on the ground.

"You two alright?" Walt raced toward them.

"We're fine. I think. Dan?"

"Take more than a couple bullets to scare this old buzzard." Dan stood up slowly and brushed the dust from his hands. "Thanks, Boss."

"You bet." Kevin looked over at the body prostrate on the ground. "Shit! Waterman's dead; we won't be able to find out who else is involved."

"I still think Sam's part of his little party. Let's make him talk."

"Thanks, Dan. But if you don't go home to Betty, I'm going to make you retire early. I thought I told you to stay out of here this weekend."

"Oh, hell, Boss. Where else am I goin' to go? I been comin' out here so long I don't even have to steer the truck. It just bring's itself."

"Well you're finished for tonight. Go home and have a beer. We'll find Sam."

"Let's get back to the shop and see who's around."

Walt gave Kevin and Dan a ride back to the melt shop. Kevin escorted Dan directly to his truck. "Go home, or I'll call Betty and have her come and get you."

"That's dirty, Boss. I almost died for you, and this is how you repay me."

"Yes. I'm sending you home safe and sound."

Walt and Kevin watched Dan's pick-up pull out of the parking lot. "He's a good man."

"Yeah, I know. Sam's truck isn't in the lot. He's not supposed to be here until Monday morning."

"Get me his address. We'll go to his house. I'd rather find him at home anyway. There's to many places to run out here. What about the scrap side of this mess? Any idea who might be Waterman's supplier?"

Kevin explained his suspicions about Anthony Mason and Adam. "I hate to admit it, but Adam was in Colorado. He's always had a thing for Laura. We've had some good times, but being out there was too great of a coincidence."

"But you have no real proof of anything?"

"Nothing."

"Either of them particularly close with Waterman?"

"Not really. Adam was closer to me and Anthony to the President of the company."

"Could Waterman have pulled this off without a supplier knowingly being a part of it?"

"I guess. Nobody really questioned his decisions. The President had complete confidence in Waterman's ability to operate the place. Waterman was smart."

"He was greedy too. That was his flaw."

"No. He had a ton of flaws. Greed wasn't even the worst."

"We'll check out these two scrap dealers, but if we don't find somebody else who knows what's going on, the game may end here."

"At least the worst is over."

CHAPTER NINETEEN

"I can't believe Daddy sent us shopping. Are you sure he wasn't suffering from some mental disorder?"

"I don't think he wanted me sitting around the house dwelling on the past few days. I'm surprised he convinced you to meet us in Birmingham for a Globetrotters."

"Oh, I felt sorry for him. With Megan getting married so suddenly, he feels his daughters have deserted him."

"So he made you feel guilty for what she did?"

"Basically, but then he promised to pay my shopping bill. A little guilt for a few new clothes: I'm getting the best part of the deal. Where are we going?"

"Don't be silly. Dad's paying the bill; we're going to Caché."

"My favorite. What'd Meg have to say this morning?"

"Luke's feeling better, and they're leaving for A&M today. Here's a good parking place." Laura pulled the blue Crown Victoria into the open space.

"Why didn't Daddy come with us?"

"Honey, there's only one thing your father dislikes more than skiing and that's shopping."

Krista laughed as she closed the car door. "He'd rather donate body parts. I'm still surprised he's letting you out of his sight."

Laura hadn't heard a word. She studied the length of the parking lot scrutinizing each person who came into her line of vision.

"Mom, what's wrong? Mom!"

"What?" Laura casually looked back to her daughter.

"Is there something wrong?"

"No. I just have an uneasy feeling. I must be paranoid."

Steel Illusions

"You have reason to be wary. I'll never be as comfortable skiing as I have been in the past. I guess that's the byproduct of being a victim."

"I know your dad feels the danger is past, or he wouldn't have let us out alone today, but I'm not so sure. There's still so many unanswered questions. Like, what was Rick doing at our house."

"Maybe he can find the answers today. Wasn't he going to meet Dan and the police detective the meltshop?"

"Yes. They're still looking for Sam. He wasn't at home last night."

"Does he work on Sunday? Why would he be at the shop today?"

"They suspect he may be there against his will, like dead against his will."

"Another one? Gee, no wonder you're looking over your shoulder."

"How's Brad today?"

"Much better. He think's he'll be released tomorrow. He can't afford to miss school any more than I can. I'm more worried about Chelli."

"Why? She only had a minor concussion. Megan said she was being released this morning."

"She's been released. I talked to her. Her father is on his way to get her and bring her home."

"So what's the problem?"

"She told him she wasn't coming back."

"She wants to stay with Stewart?"

"Yeah, her father's furious. I'm surprised Daddy took Meg's marriage so well."

"What could he do? Support her or loose her. You're both old enough to make your own decisions whether we agree with you or not."

"Yeah, but can you imagine his storming out there to drag her back?"

"Don't kid yourself. He'd go in a heart beat if thought he could change her mind. He thinks he's lost his baby. You won't understand until your own daughter gets married. C'mon, let's go buy sexy dresses for the banquet." Laura led the way as she and Krista planned to enjoy their mission of the afternoon: spend Dad's money. Laura never saw the man sitting five cars down from them, but his eyes never flickered from her.

"Sounds good to me." Krista answered. Just as Laura was about to turn around, Krista linked her arm through her mother's and led her toward the mall entrance.

****

Kevin met Walt at the meltshop with the intention of searching Sam's office. The shop was looking better. Operation of the good furnace had begun without incident. Dan was back on the job bright and early in spite of Kevin's protest.

"Boss, you gotta get this legal mumbo-jumbo cleaned up, and I got steel to make."

Kevin gave in easily. Dan was trying to give him the best support he could which was running the meltshop while Kevin took care of the details surrounding the explosion.

"How's Rick?" Kevin asked when Walt came within hearing distance.

"Better. Doctor even thinks he might make it. Did Rick tell you anything else last night? Like why Waterman would've shot him?

Steel Illusions

We were so busy last night I didn't realize how strange it was for Rick to be sitting on your patio until I got back to the office and started going over the facts with one of the officers. Do you know why he was there? Why the patio? Why didn't he wait in his car if he knew you weren't home? By the way, I did check the airlines. I'm glad you were on the flight."

"What? You didn't believe me? Thanks for the vote of trust!"

"It's my job to check, and I've known Rick for a long time; I've only known you a few days."

Kevin understood the rationale, but he still felt uncomfortable knowing Walt was checking on him. *What a mess!* "I've got my own set of questions. Like why was he shot at my house? How was he shot in the back while he was sitting in a chair, and how did Waterman know about the report? Why did he go after Dan? We know Dan saw him in my office, but how did Waterman find out? And there's still the matter of Pete. This won't be over for me until I find out what that kid was doing down by the furnace." Kevin had forgotten about the rest of the broken conversation with Rick. "Shit! I completely forgot. He said the report from his friend was done. He told me to call you. Did he give the report to you?"

"No, but that might be why Waterman shot him. I still don't know how Waterman found out about the report, but we found this in his coat pocket when we searched his body. I can answer one question for you. We don't think Rick was shot at your place. When we went over his car, there was blood on the seat. I figure he was shot somewhere else and drove to your house. It was dark; I'm only guessing that he assumed you'd have a better chance of finding him if he wasn't in his car. Did you notice it when you pulled in?"

"No. I don't recall. We were so tired."

"If he would have fallen over in the seat, he'd be dead for sure. He must've gone around to the back to find the most likely place to be seen."

"Maybe. He knows we go in and out the patio door more than the others. I feel better knowing Waterman didn't shoot him at my place. It makes a lot more sense that Rick was shot and then went to the house. I couldn't figure out how he got shot in the back while he was sitting down. There was no bullet hole in the chair. I even thought Waterman might have shot him, and then put him there just to warn me."

Walt looked at the paper again. "It's Rick's writing. The official report will probably be mailed." He handed Kevin the piece of paper revealing the location of a propane tank retailer in Sand Springs.

"How in the hell did Waterman find out Rick had the report?" Kevin asked the question again as he combed his fingers through his hair.

"In this job, I've learned I don't always get my questions answered especially when one of the key players is dead. In this case almost anybody who can give us any information is dead or unconscious."

"Okay, what about this question? How can some test prove where the cylinders were purchased?"

"You'd be amazed what a forensic lab can do. Special chemicals can be used to show the serial numbers which in turn tell us where the cylinder was purchased. Dangerous containers like this have to be marked. According to this report, there were at least six different cylinders."

"Let's go find out who bought them. There's got to be some sales records."

"As long as we're here, we might as well search Sam's office first." Kevin filed through the desk drawers. Eventually he came to a

Steel Illusions

file with several loose pages in it. He recognized bits of original reports mixed with numbers scratched out and rewritten. It was obvious Sam had been fixing the information before he turned it in. There was no sign of Sam.

"So we have one more piece of the puzzle. Sam must'a been his partner. When Waterman was shot last night, Sam split."

"No, that doesn't fit Waterman. Sam was a local boy. Waterman was too arrogant to be partners with somebody like Sam. He used Sam. I think when everything started to fall apart, he killed Sam to keep him from talking."

"But we won't know until we find a body."

"Maybe we should have the guys sift through the scrap piles. That seems to be his favorite killing spot."

"Good idea. I'll get another detective out here to oversee it while you and I go find out who bought those cylinders." Walt led the way to the car.

The drive to the small shop where the cylinders had been purchased was short. The place catered to the weekender who enjoyed a Sunday afternoon barbecue but had forgotten to fill the propane tank for the grill. The owner, an older man with a large wad of chewing tobacco bulging in his jaw, was glad to help them. It'd been a slow afternoon, and March wasn't exactly known as the barbecue season. The two week old sales receipts had been filed in the back, but were easy to locate. Kevin handed half the pile to Walt and started to flip through the small pieces of paper.

"What ya lookin' for? Maybe I can help ya."

Kevin had gleaned through several of the receipts before realizing the man would probably remember a large purchase. "Sir, do you remember somebody coming in about two weeks ago and buying an unusual number of tanks?"

He rubbed his chin as if the action would start the gears of his brain rolling. "Yeah, I think I do. It was weird. When I suggested it'd be cheaper if he bought one big one, he said to mind my own business."

"Keep looking, Kevin. We need the receipt too. Any idea who the guy was?"

"Nope. Didn't ask. Didn't want to know."

"Was he a big, tall guy with black hair?"

"Nope." The response was quick as the owner puckered-up and spit out his tobacco juice. "He was tall, but real skinny, blond hair."

"Sam. Here it is." Kevin pulled the white piece of out of the pile.

Walt laid the larger paper with the serial numbers on the counter. There were two full sequences and four incomplete. "It would've been impossible to track this place down if we hadn't had at least one full number. Let's make sure the numbers match."

"Here's one." Kevin pointed to the third series of digits.

"A couple of the partial ones, here and here." Walt put his finger on the numbers he'd picked out. "That's funny, the other full number isn't here."

"Maybe it didn't come from this store."

"Yeah, records show it did. Let's keep looking."

Kevin flipped through several more receipts before finding one that might be a match. "What about this one?" He handed the paper to Walt.

"Yeah, it's a match. I wonder why it was bought separately."

The owner answered the question. "Different guy. A local, comes in once-in-a-while. His kid died awhile back."

"Pete? I think it's time I paid Amy a visit. No wonder Dan said she wasn't talking."

Steel Illusions

"Kevin, how would Pete get the tank into the furnace? He couldn't put it in his pocket?"

"I don't really know. Like you said, we may never find out. The only one who can answer the question is dead. He either put it directly into the furnace before the other scrap was dumped in, or he added it to the scrap before it was delivered to the furnace. He could've put it in a big bag of some kind. Camouflaged it so no one suspected what it might be. None of the guys would've suspected Pete was walking around with a propane tank."

"Thanks, Sir. May we take these two receipts?"

"Sure."

****

Amy had seen visitors all afternoon on Saturday and was glad to have Sunday at the family home to collect herself. She'd always found the rolling hills in the Keystone Dam area soothing. The vegetation was just beginning to show new signs of life. Small green buds were sticking their heads out to wave a greeting to Mother Nature. The eagles, which nested in the area along the Arkansas River just below the Dam, were more easily observed before the vegetation took on its prolific summer form. Amy watched the magnificent creature soar above the tree tops. *What am I going to do? Pete's heart was in the right place, but he was wrong.*

When Amy's mother came into the room to announce Kevin Bradford was on his way over. Amy was almost relieved to find out it was Bradford rather than Waterman.

"I have to tell him. You understand, don't you?"

"Yes. I knew you'd make the right decision."

"Pete was wrong to try and swindle money out of the company I can't sue for something he planned. Even before I knew he'd

R. Z. Crompton

planned it, I felt guilty about what I was thinking. As long as I focused on my anger, I didn't have to face the truth. We always knew working at a steel mill was dangerous. He told me many times how unpredictable things could be out there when the electrodes were bearing down. I can't believe he thought he could control the situation. Oh, well. I can't shake him now, but I'd like to. Let me freshen up, Mom. Call me when they get here." Amy combed her hair and put on some light pink lipstick. She buttoned her blouse just as she heard the door bell. She felt more at peace with herself and Pete's death now that she'd givin up her anger.

Amy's mother looked through the door and announced, "There's a police officer with him."

"He knows then."

"Yes, Dear. I suppose he does."

"Well it's best to get this taken care of before the funeral. Will you go out with me?" The older woman shook her head affirmatively. Amy picked the letter up off her dresser and put it in her pocket, then left the room with her shoulders squared ready to face the truth. Her hand reached down to touch her stomach. *Your Daddy loved you very much. Don't worry little one. Grandma and I will take care of you.*

Kevin Bradford was standing at the large picture window when Amy entered the room. The other man had taken a seat on the over stuffed sofa. Amy look around the soft, powder blue room and remembered back to her days of playing dolls here. It was good to be moving back. This place held good memories for her. Pete would rather have her bring his child here instead of to the apartment he hated. There had been only sadness in the small set of rooms. It wasn't a place she could go back to.

"Hello, Mr. Bradford."

"Hello, Amy. My deepest regrets about Pete."

Steel Illusions

She clinched her jaw to keep control of the tears and motioned for him to be seated. "Thank you. I'm sorry about the other night at the hospital." She was ashamed of what she'd said.

"It's forgotten. I'm sorry to bother you, but we need to talk to you about what happened. Do you know anything at all about what Pete was thinking before he went to work Wednesday?"

"I'm afraid so. He was agitated and wouldn't tell me why." She pulled out both pieces of paper Pete had folded and placed in the envelope. "He mailed these to me. I found them in the mailbox last night."

Kevin looked at both pages before handing them to Walt. "The night letter wasn't from me. It's got my initials on it, but I didn't write it. My guess is that Waterman sent it."

Amy looked confused. "I don't understand what he has to do with this. I only know Pete had something planned that wasn't good."

"Amy, he bought a propane tank the day of the explosion. Must'a been on his lunch break."

"Oh, My God. I didn't think he'd gone that far."

"It's ironic that Sam bought the others from the same place." Walt said.

"Not really. It's the only place in town. Sam was stupid for not going to Tulsa or even Oklahoma City to get the tanks he wanted. Actually, he should've gone out of state."

"Waterman should'a taken care of that himself."

"Excuse me, what does Sam have to do with my husband?"

"Amy, Waterman's dead? He was embezzling from the company and wanted to make it look like I was the guilty one."

"Waterman's dead. That's terrible. But Pete never had any money. We were broke."

"I know. Waterman used Pete. The bad propane tanks were already planted in the scrap. Pete just planned a small explosion of his

-310-

own. He had this night letter giving him reason to be on the floor. No coat meant he'd probably have some burns, but nothing life threatening. What he did was wrong, and he paid dearly for it. I'm sorry."

"So am I." She could hardly get the words out. "I knew..." she choked on the urge to cry. "...there was something wrong when Mr. Waterman kept saying he wanted to help me get even with you. It just didn't seem like the thing a V.P. would do." She leaned back in the chair and wiped the tears from her face.

"Amy, Pete was set up more than he was guilty. I don't think he would have ever done anything like this without having the opportunity thrown in his face."

"Thank you, Mr Bradford for saying that. Will you be at the funeral tomorrow."

"Yes, we'll all be there. May we take this?" Kevin held up the night letter, and handed her back the profession of love from her husband.

"Don't you need this?"

"You keep it. It's private."

"Thank you. This means a lot to me." She reached up and kissed Kevin softly on his cheek before tenderly folding the piece of paper; the last words Pete ever wrote to her.

Kevin and Walt excused themselves and closed the door behind them. "What are you thinking?" Walt snapped at him. "The kid is guilty too, at least in the part he played."

"He's dead. He's paid for his sin. There's no reason to make her suffer any more than she already is. Pete was desperate; Waterman was greedy. There's a hell of a difference in my book. I don't want anything coming out that'll keep her from getting the company settlement."

"I didn't hear all this."

Steel Illusions

"Thanks, Walt. Where to now?"

"The hospital. We only need to find out why Waterman confronted Rick, and most of our pieces will be in place."

****

Laura and Krista entered the elegant dress shop still arm in arm, talking wildly about the clothes they were going to take with them to Birmingham. The lady behind the counter heard them coming.

"Laura, is that you?"

"Denise, it's so good to see you." Laura hugged the petit, young woman as she came around the counter.

"We were so worried about you. You're okay?"

"Only these bandages." Laura held up her wrists. Her long sleeves had hidden the white gauze from easy observation. It had occurred to Laura that people would think she tried to slit her wrists.

"Is it bad?"

"Only the itching. I put on clean bandages this morning, and the rope burns are looking better and starting to itch."

"When did you get back? The news reported this morning that you'd been found only yesterday. I can't believe you're here. Actually, I'm even more surprised Kevin let you out!"

"Laura? It's you!" Angela came out of the back room carrying several pieces of new spring fashion. She tossed the clothes on the counter and threw her arms around Laura. "I was going to call you this week to see how you were doing. I had no idea you'd be back so soon."

"Mother, how much shopping do you really do here?"

The three women glanced at each other and back at Krista. "Enough, just enough," was the only answer Laura gave her daughter.

"Krista, we haven't seen you for a long time. How's school?"

"Hard. I don't have much time to go out, but since Dad said he'd pay for the trip, I figured I'd better take him up on the offer."

"What trip? You guys are leaving town already?"

"Yup. And we need to be dressed for the event."

"What's the occasion?"

"A Globetrotters." Laura announced.

That was all Angela needed to know as she led Laura and Krista around to inspect the new spring selections. "Let me put these in a fitting room for you. Was Kevin sick? He never lets you out to shop on a Sunday."

"He must'a been. He even said he was paying for all of my new clothes too." Krista was still not sure she'd heard her father correctly. Repeating the fact seemed to make it more real. "Ya know, Mom, I don't think Dad has any idea how damaging it can be to send the two of us out with unlimited approval."

"Yes, he does.

"Let me bring you a smaller size, Laura. You've lost weight."

"Thanks, Angela.

"Mom, you have lost weight. Are you sure you're okay?"

"I'm fine. I've got a doctor's appointment tomorrow just to get checked out, but I feel fine."

"Are you sure? Now that I think about it, you look really pale. Is your appointment tomorrow with the plastic surgeon or your regular doctor?"

"The plastic surgeon."

"Mother, he can't give you a general physical. When are you going to your doctor?"

"As soon as I can get an appointment. Ya know, it's hard to make appointments on Sunday. I appreciate your concern, but give me a break. I'm seeing the plastic surgeon first because I'm dying to have more than a sponge bath. I'll take care of the other as soon as I can."

Steel Illusions
___

"Before Birmingham?"

"Yes. Now stop giving me a hard time. This is supposed to be fun. We're shopping, not at a mother - daughter inquisition."

"I'm just worried. Dad has been a little preoccupied since we got home with bodies on the patio and all, and you've been very vague about what happened while you were at the cabin."

"There's nothing to tell. I was tied up and drugged. The next thing I knew Megan and Luke were there to rescue me."

"Ah, right. And I was born yesterday. This may not be the place to talk about it, but I was part of that rescue, and the song you are singing doesn't match the damage to your arms."

"You're right. This isn't the place. I'm fine. Considering the hike I made, I should've lost a couple of pounds. So I'm a little pale. Who wouldn't be? I'm glad to be alive. Now zip me up. What do you think?" Krista zipped the black lycra up the back. Bright yellow, pink, and purple suede squares made up the front. The skirt was a black pleated chiffon, sheer to the waist. It was simply stunning.

"It's fantastic. You're drop dead gorgeous. Daddy will love it."

"Thanks, Honey."

"Laura, how about something like this?" The beautiful, young woman brought Laura several other pieces to choose from; silk, linen and cotton. It was wonderful being pampered after the brutalness of the last few days. "Can I get you something to drink while you change?"

"Thanks, Angela. This turquoise and white sarong is wonderful. I don't think it will be warm enough to wear it in Birmingham, but it'll be great for this summer."

"Mom, how do I look?" Krista came out of her dressing room in a shocking evening gown. "I love this dress." She ran her hand down the sexy black material. It was floor length, but it wasn't. The sheer nylon mesh flowed to her ankles, but only a small amount of her body was actually covered with a black lycra which accentuated every part of her slim well proportioned body. Small silver orbs studded the shoulders and arms.

Laura was speechless as she watched her daughter parade in front of the mirror.

"What do you think? Lycra is really popular this year."

"I guess I never considered it evening wear, but the dress is certainly amazing."

"Do you think Daddy will like it?"

"He'll like it, but I don't think he'll let you out of the house."

"He's not going to see it until we get there. I'm not a complete idiot. What else do I need?"

Krista was like a wild woman as she went from rack to rack choosing a variety of styles. She modeled everything with Laura giving her the final approval. Laura had more fun watching Krista go through the styles than she did changing herself. It was difficult getting the items on and off with the bandages at her wrists and ankles. And she was tired, more tired than she'd realized. After picking a few things for herself, it was easier to sit on the sofa and watch Krista parade around.

Angela brought Laura another glass of water while Krista finished trying on anything that caught her attention. Several other customers had been in and out of the store by the time Krista was finished.

Matching accessories were found for each of the outfits before Laura was able to get Krista out of the shop. "Thanks for everything, Angela."

"You're welcome. Laura why don't you wait here while Krista brings the car around. I'll help her take everything out."

"That's a good idea, Mom. You really don't look well."

"I can make it to the car. I'm just tired." Krista and Angela looked at her. "Really?"

"Maybe your body is still reacting to that drug he gave you."

"Who knows. But I would like to get home and rest."

Krista finally convinced her mother to let her go get the car and drive home. Angela helped carry the bags out and gave Laura a warm hug before closing the door for her.

"Have fun in Birmingham."

"Thanks, we will."

Angela watched the car pull out of its space but had turned to go back inside before the large black sedan pulled up behind the Crown Victoria and followed it out of the parking lot.

CHAPTER TWENTY

The opulent Wnyfrey Hotel in Birmingham, Alabama, hummed with the business of the day. Hundreds of guests related to the steel industry were checking in. Ornate oriental vases, many of them privately owned, decorated the lobby area where Laura sat while Kevin registered. She enjoyed the fresh exotic flowers in the immense porcelain container. Large columns covered in leather blocked her view of the elevators, but she recognized the constant tone of the elevator signaling its arrival.

Laura had spent the five days prior to their flight resting and looking over her shoulder. Krista left for school Sunday evening after Kevin returned from his visit with Rick. She was more interested in her mother's welfare than her father's successful day of sleuthing. She almost didn't return to school. Only after Kevin's promise that he'd make sure Laura was well cared for did she go ahead and pack her bags. Kevin appreciated Krista's concern for her mother, but he felt she exaggerated the seriousness of Laura's condition. After all, Laura was entitled to have a few shadows after what she'd been through.

Laura slept peacefully for about six hours and then woke in a cold sweat. Kevin was sleeping soundly beside her; he never felt her get out of bed. Krista was already gone, so her only companion was the dog. He followed her from window to window as she scrutinized the darkness for something; anything that shouldn't be there.

Laura had trouble sleeping all week. She was either totally exhausted and passed out only to wake terrified, or she paced the floor for hours. Shadows followed her everywhere. Her only escape seemed to be in books. She read everything she could find from six month old magazines to the hottest best-selling novels. She had to concentrate on something besides the ordeal in the mountains. The doctor had given her a clean bill of health physically. They had talked about the details of the abduction, and Laura seemed to handle the reality of what had happened fairly well. Her doctor understood her rationale for keeping her secret personal and told Laura to call her if she needed more help. It wasn't the rape which haunted Laura; it was the feeling that the trouble wasn't over.

Kevin tried to cancel the trip, but Laura wouldn't hear of it. She was obsessed with getting out of town. Once the plane had taken off, she slept peacefully with her head on Kevin's shoulder. He was concerned more about her appearance than her mental state, but then he'd never asked her why she seemed so unsettled. He didn't want to make her relive the tribulation of the kidnapping; however, she looked thin; too thin. And he was sure that the healthy glow in her cheeks was due to a good make-up job. Kevin honestly believed her physical and mental health would improve with time and care.

"I've got the keys, Dear. Let's get you upstairs."

"Okay. When does Krista's flight get in?"

"She said she'd be here by ten o'clock. We'll be back from dinner by then."

Kevin led the way to the elevator. As he turned the corner, he came face to face with Adam. The two men hadn't talked since Adam's return from the Rockies.

"Hey, Kevin. It's good to see you, Buddy."

Kevin's greeting was cool. "Hi. Where's Stephanie?"

"At the last minute, she had to cancel. Susie came down with the chicken pox."

"Laura, you're looking wonderful." Adam reached over to give her a kiss on the cheek and a hug.

Laura managed to stand still while he finished his greeting. "Hello, Adam. Sorry about Steph not being able to make it. Please excuse me. I need to get upstairs."

R. Z. Crompton

"Will I see you later?"

"Probably."

"Kevin, you wanna go have a quick beer and see who's checked-in?"

"I'll come back down. I want to make sure Laura gets settled in."

Laura didn't say anything as the elevator made its way up to the club level. She seemed agitated as the door opened and she looked into the hallway.

"Are you okay, Dear?"

"Yes. I just need a few minutes to freshen up and unpack then we can go back downstairs. I enjoy watching the people come in."

"I'd feel better if you rested for a while, or you won't make it through dinner."

"I'll rest for a while if you'll go downstairs and have some fun. Everybody is going to want to talk to you about Waterman and what's happened. You're lucky they aren't up here knocking on the door."

"Can't have fun just yet. The committee is meeting in a few minutes. Are you sure you'll be okay?"

"Yes. You can't babysit me forever. I'm fine. I'll read for a while. If I sleep, I sleep."

Kevin helped unpack and filled the ice bucket, so Laura could have something cold to drink. After he changed into something more comfortable, he kissed Laura good-bye and left the room.

Laura took her light green and black tapestry styled jacket off and hung it in the closet. The knock at the door startled her. She assumed it was Kevin. *He must have forgotten something.* She never asked who was there. Adam greeted her when she opened the door.

"Laura, may I talk to you?"

"Why?"

"Please let me in."

Steel Illusions

She was hesitant about letting this "friend" into her room, but she didn't have a good excuse to keep him out. Slowly Laura stepped back opening the door far enough for Adam to walk in.

Adam walked past her toward the burgundy sofa on the other side of the pale green room. His back was toward her until he heard the door close. Slowly he turned to confront her. "The reception downstairs was a little cool. Don't you think?"

"I guess you surprised me."

"Laura, why in the world would I want to kidnap you?"

"Shit, Adam. Why don't you just come straight to the point?"

"Sorry to be so blunt, but the police are asking all kinds of questions. Does this have anything to do with what happened when we were in the hottub? I can't believe you'd blow it so out of proportion. Can you imagine how Stephanie feels? That's why she's not here."

Laura felt guilty now. "Adam, don't sound so surprised. Kevin's sure a scrap dealer was involved, and he narrowed it down to two possibilities. I'm sorry you happen to be one of those possibilities. Can you honestly tell me it was purely a coincidence that you were out there?"

Adam didn't answer right away. The delay was enough to suggest guilt. "Not exactly. I did have business, and then I came looking for you. Did you tell Kevin about...about us?"

"There is no 'us'. We didn't do anything."

"If you didn't say anything, why is he looking?"

"Think about it. You're one of his main suppliers. You were there at the time of the abduction. You got held up by a man-made avalanche. That's enough to raise suspicion."

"I can't believe either of you would think me capable of such a terrible thing."

"Why did you feed me the line about Kevin looking for gloves? You never had the conversation with him. I had the feeling you wanted me to check on him. Find out what was going on at the mill, so you'd know whether or not to go on with your plan."

Adam's face took on a more sheepish look as she threw the accusation at him. "I was fishing. I wanted to know if it was possible for Kevin to just show up. I know he's flown out to meet you before when you and the girls were on vacation. I wanted to enjoy you myself. That's all I'm guilty of, not planning a kidnapping or murder. I can't believe after what I said to you, you could even consider me a suspect."

Laura caught herself wondering if he had been the one to "enjoy" her. After all, she had been drugged. *No. There's no way he would've been able to hike in and out of there.* "If you're so innocent, why are you acting this way? The police haven't found anything, have they?"

"No, they haven't. There's nothing to find."

"Then relax. Kevin and I don't want you to be guilty."

"Then why point the finger? Why didn't you stick up for me? I'm not concerned about what the police might find. I'm worried about what you think."

"My God, Adam. It's a police investigation. They asked the questions. We answered. Can you even begin to imagine what this has been like for me?" Laura yelled at him. Her patience was slipping quickly, and her pent up emotion and fear were taking its place.

Adam was surprised by her reaction. He'd never seen her lose control. "I'm...." Before he could finish his thought, she was talking again.

"How would you feel if Stephanie had been dragged off into the wilderness?"

"I don't have to wonder how I'd feel. I would feel the same as I did when I found out you'd been taken."

Laura didn't respond. She let the words sink in. "What are you saying?"

"You know exactly what I mean. I understand how Kevin feels. What I would like to know is if you're so afraid of me, why are you here?"

"I'm not afraid of you."

"Then what? The police have gone over everything in my office and my house. They've covered bank records and even hospital files. They have nothing. Laura, look at me."

She didn't turn around. Adam walked over and grabbed her by the wrist. Laura winced at the pain. "What in the hell is wrong with you. Do you really think I'd hurt you?"

She held her wrist tenderly as she turned to face him. Her pasty white complexion caught him by surprise. "What's wrong with you?"

"Nothing." She was hardly able to get the answer out.

"Like hell. What happened to you up there?"

This time there was no answer, only a negative motion of her head. Even Kevin hadn't drilled her about the abduction. He taken her word for it that she was okay. And she was determined to be "okay".

"Show me your wrists."

"I can't. They're bandaged."

"Both of them are injured? I want to see."

"I can't."

"You mean you won't. Have you told Kevin the truth about what happened to you?"

"What's that supposed to mean? Of course, I did."

"Yeah, I bet. Did you show him your wrists?"

"No, he didn't ask. He's been just a little preoccupied with all the Waterman mess. You know being shot at and finding bodies tends to occupy your thoughts." She answered sarcastically.

"Well, I'm not occupied and I'm asking. Now take off the bandages or I'll do it myself."

"No, Adam. You don't have any right to ask or see."

"I sure as hell have a right to see what I'm being accused of. You're terrified, aren't you? You still think somebody's after you. Does Kevin know how you feel?"

"This is none of your business."

"I'm making it my business. Now show me." He demanded.

Laura had backed herself up against the bed. There was no where to go. The wounds were looking better everyday, so what did it matter if he saw them? It was better to show him herself rather than to wrestle over something so silly. She sat down on the bed and pushed up the sleeves to the silver knit shirt. Slowly she unwrapped the bandage she'd put on early in the day. Then she held out her arm for inspection.

"Jesus Christ, Laura. It doesn't take a rocket scientist to figure out what happened. This is after a week of healing? These aren't simple rope burns. Ankles too, I suppose?"

Laura looked at him in appreciation. His reaction was sincere and she felt better. "Adam, there's nothing to tell that would make any difference."

"I'd kill the son-of-a-bitch."

"Yes, I'm sure you would, but I'm not telling, and you're only guessing at what happened. That's the way it's going to be."

"How can you be so calm about this and terrified at the same time?"

"I can deal with what happened. Shit, I was drugged most of the time; so the memory is a little foggy to say the least. She lied about not remembering. Jake's face was branded into her memory. What I can't shake is the feeling somebody is watching me all the time. That's why I came with Kevin. I wanted to get out of town. I was hoping to get some rest."

"I know the other guy who's being checked out. Have they found anything to tie him to this kidnapping and..." He hesitated using the four letter word. It was more repelling to him then any other word or act he could conceive.

Laura got them both off the hook by finishing the sentence for him. "...and murder?"

"Yeah. Close enough."

"Nothing. At this point, Kevin and the police are figuring Waterman was in it alone."

"So what makes you think both things are tied together?"

"I heard the kidnappers say Kevin's name. They wanted him out of the shop."

"I take it both of the kidnappers got away?"

"In a manner of speaking. One was dead, but the injured guy carried him off. So there's absolutely no evidence that the two incidents are related, only what I heard and our gut feeling. Luke said the dead guy's I.D. had a Tulsa address, but he didn't remember the name. He didn't expect the body to disappear."

"Why didn't he put the I.D. in his pocket or something?"

"I don't know. The kids were more concerned about me than a dead man. The funny thing is the more I think about the dead guy, the more I think I knew him from somewhere."

"Did you come here hoping to get away or hoping to confront?"

"I don't know. I really don't know."

"Have you told Kevin about your fear?"

"No. He would've stayed home. I can't live locked away in my house. I won't turn my home into a prison. I've been coming to these meetings for over ten years, and I'm not stopping now just because I'm still getting goose bumps. It's been less than a week since I was whisked away into the wilderness. I'm entitled to a few goose bumps."

"Maybe."

"I'm sorry about Steph. Does she hate us?"

"No. She was afraid you'd hate her."

"So you didn't tell her about the hot tub incident?"

"Didn't seem like the smart thing to do."

"So what do we do now?"

"You put your bandages back on, so we can go downstairs for a drink. Between me and Kevin, you will not be alone all weekend."

"He's too busy to worry about my shadows."

Adam studied her demeanor. "You don't want him to know."

"His part of this is over. All the pieces fit. When he went to the hospital last Sunday, Rick was well enough to fill in the last few gaps."

"So tell me what he found out."

"Waterman confronted Rick at his home late Saturday afternoon. Rick was waiting for a call from his friend with information about the explosion. They argued. Waterman wanted to know why Rick was snooping around in something that was none of his business, and then he threatened him. Waterman left without ever revealing a gun. After Rick got the call he was waiting for, he left the house and headed for our place. He'd called to find out when the last flight from Colorado Springs would be in and planned to be there waiting for us. When he parked his car in front of our house, he got out of the car and started for the back of the house. He was shot in the back. He fell to the ground, but managed to get back to the car. After a few moments, he realized he couldn't drive himself anywhere; and he was afraid we wouldn't find him, so he crawled out of the car and managed to get to our back door."

Steel Illusions

"Why didn't he go to a neighbor?"

"There were no lights on anywhere, and he knew he only had so much time to get himself to a spot where we could find him."

"I thought Kevin was told Waterman had shot Rick somewhere else."

"That's what we originally thought because of the blood in the car. But only Rick could tell us exactly what happened. Maybe that's one of the things that bothers me. We took the evidence we had in front of us and came up with a logical explanation, but it was wrong. It was still wrong." Laura shook her head.

"Didn't Rick have a phone on him? I can't believe a reporter wouldn't have a telephone in his pocket."

"Kevin asked the same thing. Rick always let Kevin use his cellular phone; but after he left the message for Kevin, he didn't have the strength to make another call. The other big question he answered for Kevin was about Dan. Kevin couldn't figure out why Waterman went after Dan. Rick said Waterman admitted he was taking care of the loose ends. Sam, Dan and Rick were considered the loose ends. Then he was going to make it look like Kevin had done the whole thing."

"Sam, who the hell is Sam?"

"The only guy Kevin was able to tie to Waterman. Kevin thinks there were others involved, but Sam had the most direct tie. Dan got sucked in because he confronted Sam by himself Saturday afternoon. Dan's just lucky that Rick was conscious enough to tell Kevin a few pieces when we got home; otherwise, we would've come home and gone to bed. How did you find out so much about all this?"

"I do have a man who works on-site. Remember?"

"What does he say about all this?"

"He thinks it's Anthony Mason, but then he's supposed to be on my side. I make out his paycheck. Did you suspect me only because I was in Colorado at the wrong time?"

"Yeah, if you hadn't been there that night, I doubt Kevin would've even mentioned your name. I think he got to the point where he couldn't assume anyone was free of suspicion."

"Just because I was thinking with my lower brain, as Megan would've put it, I nearly destroyed our friendship."

"Which brain are you thinking with now?"

"After all this trouble, I had a 'lower brainectomy'; now I can only think with the brain on my shoulders." Adam let out a hearty, wholesome laugh.

Laura shared the light hearted play with words. "I bet Steph is disappointed."

"Oh, I didn't say the lower brain doesn't work; it just doesn't do the thinking any more."

"I'll believe that when you prove it."

"So Kevin thinks this mess is pretty much over?"

"Seems to be." Laura shrugged her shoulders in a sense of frustration.

"But not for you?"

"I don't know." Laura got up off the bed and walked across the room rubbing her hands together as a sign of her anxiety. "All the questions surrounding the mill have been answered. But none of the answers tie the kidnapping to the explosion. We're still speculating. Maybe it's the speculation rather than concrete answers that's bothering me. Like I said before, we looked at all the facts; but still came up with the wrong answer."

"I still think you should tell Kevin what's going on. I learned a long time ago to trusts a woman's intuition, and I don't think you should be alone."

"You are kidding, right? I don't want somebody in my room with me every minute."

"Why not? You're in as much danger here, maybe more. It's not like you won't open your door to just anybody."

Laura understood he was referring to the fact she'd opened the door for him without asking who was there. "I thought you were Kevin."

"And you were wrong. What if I was the bad guy?"

"I see your point, but Krista will be here tonight. She can stay with me. I don't usually spend that much time alone anyway. I don't come to the Globetrotters to sit in my room."

"This is different. I still think you should tell Kevin."

"Tell him what? That I'm scared of my own shadow?"

"Yes. You're his wife. If you're afraid of something, he should know."

"I'll think about it." Laura finished replacing the bandages and grabbed her purse. "Let's go downstairs."

Adam opened the door for her, and the two of them headed for the elevator. They never noticed the person standing at the ice machine.

****

Laura approached the lounge area feeling better knowing Adam took her seriously. She'd not wanted to tell Kevin because he'd have explained her shadows away with silly excuses. He believed she just needed time to forget. Kevin didn't believe in hunches or intuition. He wanted facts; something tangible. That's why it'd been so easy for him to believe the case was over. All his questions were answered, and there was no evidence that Adam or Anthony were involved.

R. Z. Crompton

"Laura! Laura, come and have a drink with us." Charlie, nicknamed by the ladies as Mr. Stud Salesman, greeted Laura with a bear hug which took her breath away.

"Hey, Charlie. It's good to see you."

"Tomorrow I'm hosting a tour of my fair city. Will you join Pat and me."

"I'd love to. Will you have room for Krista?"

"Another beautiful woman. Always. What can I get you to drink?"

Laura appreciated Charlie's tact in not asking her about being kidnapped. She was sure it was the first topic everyone wanted to talk about. The concern was genuine, but she didn't want to relive it several dozen times over the weekend. In fact she was hoping if she stopped talking about the ordeal, maybe the shadows would go away. This trip might be the therapy she needed, and it was sure a lot cheaper than a doctor.

"How about a glass of champagne, Charlie? It's really good to be here."

He gave her another kiss on the cheek and whispered, "Pat and I are really glad you decided to come."

The cocktail hour was a constant reunion as each newly arrived member appeared at the door. This was just the diversion Laura needed. She spent hours talking with old friends about jobs, children and past Globetrotters. For the first time in a week, she went for more than a few minutes without thinking back to the days in the mountains. Kevin was surprised to find Laura at the bar listening intently to the newest series of jokes being exchanged. Tradition reigned as steelmaker tried to top steelmaker with one joke after another. Kevin enjoyed watching Laura relax. She didn't even jump when he walked up and put his hand on her shoulder.

"Hello, My Dear."

Steel Illusions

"Hi, Honey. Did you take care of business?"

"Sure did. You ready for dinner?"

"No, I'll go up with you so I can change." Laura got up from the chair and locked her arm through his. Now the glow in her cheeks was natural.

Laura tried to retell the better jokes of the afternoon as they waited for the elevator. Kevin was more interested in some extracurricular activity than jokes. Tradition was tradition, and this was the best time of the day for loving.

****

Krista was tired and irritated by the time she got to the hotel. Grand or not, registering at any hotel was a pain in the butt. Kevin had booked her a private room, which was fine with her. She was sure it was because he wanted privacy with his wife. She'd enjoy the peace and quiet of a single room a lot more than she wanted to share space with her mom and dad. Unfortunately, that meant she had to register herself. Finally, with key in hand, she turned to head for her room. In her haste, she forgot to look before she walked. Krista ran directly into a concrete wall dressed in a double breasted suit.

"Excuse me. I'm really sorry." Krista started to apologize profusely before she even looked up.

"It's okay, Miss." The voice was deep and caught Krista's attention even more than the handsome face staring down at her.

"I was in a hurry."

"I could tell."

Krista looked at the man even after he moved to the side so she could pass. "Do I know you from somewhere. You sound very familiar."

"Sound familiar? That's a new line."

"New line? Excuse me. That wasn't a come on." Krista wanted to add "You ass" but decided to hold her tongue. She knew this arrogant ass from somewhere.

"In that case, I don't think I've ever seen you before."

That was all she needed to hear. "I heard you lecture at the University."

"I've given a few lectures in my time. What were you doing listening to me lecture?"

"Well, what do you think I was doing there? I'm certainly not a groupie."

Krista turned to walk away from the man, but he stopped her. "Are you a doctor? Are you here for the conference?"

"No, I'm merely a student. Barely good enough to stand in your mighty shadow."

"Why so hostile? I don't even know you."

Krista took a deep breath and decided to start again. "Sorry, long day."

"Will you tell me your name?" The towering man looked down at her. He saw more than a sparkle in her eyes; there was an intensity he recognized, but couldn't quite place. If he'd seen her at the school, it was very likely he'd never really noticed her. During lectures and rounds he saw students not individuals. It was a fault he'd been called on many times. His bedside manner with his patients was tender and empathetic; but with students, wanting to suck the knowledge from him only to try and prove they were better, he felt insecure. The Chief of Staff had told him it was his immaturity; and hopefully, he'd grow out of it.

"Krista Bradford, second year med student. Pleased to meet you, Dr. Summerfield." She held out her hand for the official handshake. Making enemies with this man wouldn't serve her purpose at all. He could make a big difference in her acceptance into the medical circle and the University. Learning to play the politics was nearly as important as learning the medicine.

"Pleased to meet you. Now, will you answer my second question? What brings you to Birmingham? You've made it clear it's not to hear me!"

"I'm meeting my parents for a long weekend. They're here for a steelmaker's conference."

"I hope you enjoy your weekend. Maybe I'll see you around."

"Maybe." Krista walked away confused about her feelings for the arrogant man she'd seen standing on the stage in the huge lecture hall. He didn't seem quite so bad in a one-on-one conversation. *Oh well, I may have to respect the man's brain and ability, but I don't have to like him.* Krista walked to her room where a message was waiting for her to join Laura and Kevin in the lounge. Quickly deciding it'd be better to change, Krista opened her bag and pulled out one of the new outfits her father had paid for.

Dr. Summerfield sat in the lounge away from the crowd. He'd wanted a nice cognac before going up to look over his presentation. His attention was captured by the lady in red. *How in the hell did she change so quickly and get back to the bar?* Miss Bradford had changed from the uptight, med student into a very elegant lady. *Well, this is interesting.*

Krista entered the lounge and looked over the crowd. Spotting her mother dressed in the red silk oriental pant suit was instant. She was pleased to see the animation in her mother's face as she approached. The old sparkle was back in her eyes when she looked up to see Krista walking toward her.

Dr. Summerfield was stunned as he watched the slender woman in the graceful champagne colored pantsuit walk past him. Her long curls now hung well past her shoulders and swung with the rhythm of her body. This was the Miss Bradford he'd met in the lobby. *The lady in red is her mother? How incredible. Clones.* Just beyond eavesdropping range, the doctor had to be satisfied with watching.

"Hi, Honey. I'm glad you decided to join us." Laura ordered Krista a glass of wine.

"Where's Daddy?"

"He went up to bed a little while ago."

"That's typical. How are you?"

"Having a wonderful time."

It was easier for Krista to relax now that she saw the true color back in her mother's cheeks. To Krista's surprise she knew many of the people in the room. Most had been to the house one time or another for business. No wonder her parents enjoyed going to the Globetrotters so much; all of their friends were at one big party. She'd grown up with these people and yet never realized how often her parents really got to see them. This was going to be a good time.

Krista looked around the large room to see who else she might know. Adam was a few feet away from her mother and acknowledged Krista's glance with a warm smile and tilt of his glass. He was watching, but she wasn't sure what he was looking for. Lighting was low making recognition of those on the fringe difficult. She assumed that most of the others were guests at the hotel but not with the steelmakers. The large man sitting alone was unmistakable. She felt rather sorry for him but then if he wasn't such an arrogant ass maybe somebody would invite him over.

Dr. Summerfield was surprised when Krista walked over to him.

"Hello, Dr. Would you like to put away your 'stuffed shirt' and join us for a drink? I promise I won't tell anyone at the hospital I saw you out of character."

"And do you want the same protection? You don't want me to tell anyone I saw you out partying with a wild crowd in Birmingham."

Steel Illusions

"Give it your best shot, Doctor. With my reputation on campus, no one would believe you." She turned to walk away.

He was at her side before she could take two steps. "Please call me David." He intended to find out more about her reputation in the morning. A quick call to the university, and he'd get his questions answered.

Krista introduced David into the circle of friends. He was accepted because he was linked to the Bradfords. That was all the others needed to know. The jokes and stories were shared with him as if he'd always been there. Krista was impressed with the inclusive treatment given to all. As she looked around the circle of friends, there was no way of telling who'd been coming for years and who was a first-timer. The ease and warmth of the group was refreshing.

When last call was given, Adam stepped up to escort Laura to her room. Krista wasn't sure why her mother had so easily accepted the overture of protection. "I'll make sure Mom gets upstairs, Adam."

Laura was quick to understand the reality of leaving the bar. Adam was the only one who new about her suspicions, and she needed his eyes to watch for her. He'd be more objective and knew more people. She was positive her kidnapping was not only tied to the steel industry, it was tied to the people who set up the explosion and not all of them were dead.

"It's okay. I'm on the Club Floor; you have to have a special key to get up there. Besides, Adam has to go that way. You go ahead. I know you're tired."

"Are you sure?"

Laura shook her head affirmatively and got up to leave with Adam at her side. The crowd thinned out quickly as the last drinks were finished. The men would get up early for meetings so their afternoon would be free for golf. It was a tough conference life; but somehow, they managed to get through the long hard days of meetings.

golfing and drinking. The ladies had an early breakfast and set out for shopping or tours of one kind or another.

When they got to the entrance of the lounge, Adam looked back at David, who was standing behind Krista. The man wasn't budging. He was staking claim to her, and Adam doubted if Krista even realized what was going on. *Shit not only is she the spitting image of her mother, but now some stranger is sniffing around her skirt, and Kevin's upstairs in bed.*

"Laura," Adam felt the tension in Laura's arm just before he turned to face her. She was staring at the delicate wings of the sculptured herons which decorated the Mall entrance to the Hotel. "Laura, what's wrong?" He questioned softly, but he didn't need an answer. The goose bumps on her arm was all he needed to feel in order to understand she suspected she was being watched again.

"Walk slow and look around. See if you recognize anyone."

Krista had seen the change in Laura's demeanor and quickly said her 'good-nights'. "Mom, what's wrong?" She asked coming up next to her mother.

"Nothing, I just had the strange feeling somebody was watching me. It's nothing."

"Yeah, right. And fat babies don't fart." Krista said flatly. She wasn't fooled by her mother's explanation especially since Adam had made such a big deal about getting her back to the room okay.

David, standing right behind Krista, wasn't sure he'd heard her correctly, "And babies don't what?"

"Oh, sorry, David." She moved slightly to allow him into the conversation. "Now what's going on?"

"Krista, don't worry. It's just the shadows from last week."

Adam looked at Krista and back at Laura. It'd be too easy to follow the wrong one. "You two cannot go walking around here alone. You're like carbon copies of each other. I watched Krista walk into the lounge. If I hadn't known you were already at the bar, Laura, I wouldn't have been able to tell the difference."

"I'll agree with that." David threw in. "I thought your mother was you until I saw you at the door. What in the hell's going on?"

At this point, Adam felt responsible for both women, and he could use some help. "Krista, do you know this guy?" He cocked his head toward David.

"Yeah, he's a doctor at the University. Why?"

"Fine you can tell him as much about this as you want, but promise me you won't go through the halls alone."

She fumbled for something to say. She didn't want David knowing anything about her personal life; and even though she'd enjoyed his company tonight, she didn't want him to feel tied to her.

David didn't wait for her response. Adam was serious about the danger, and there was no denying the strong resemblance of the two women. "I promise."

The four of them walked through the lobby toward the elevators without seeing a single suspicious person. They whispered about the man reading the paper, and the lady coming out of the restroom. "It's not a woman." Laura offered.

"How do you know?"

"I just do."

"I've seen some women I'm afraid of." Adam commented.

"Me too." David added.

Krista was surprised by his admission. She didn't think he'd be afraid of anyone. They arrived at the elevators and waited for the doors to open. There was no one in sight.

After Krista and David left Adam to take Laura the rest of the way to her room, David was quick to ask for an explanation. Krista described the last few days of the ski trip as they walked down the hallway. Only a partial story had been given by the time she'd unlocked her door.

"You have quite a story. May I hear more of it tomorrow?"

"I don't want to bore you with the details."

"You won't. I'll be finished with my presentation by lunch time. Can I meet you somewhere?"

"In the lobby? About noon?"

"No way. Adam would ring my neck if he found out I left you waiting in the lobby. When you get back from your tour, get your dad or somebody to bring you up to the hall. You can wait for me to finish my presentation."

"Adam can't reach your neck." Krista offered the observation with a smile.

"You know what I mean."

"David, thanks."

"Good night. And I want to hear this door lock." He waited until he heard the dead bolt and the chain. Then added, "Krista, if you need me, "I'm in 986."

"Thanks."

CHAPTER TWENTY-ONE

At four-thirty in the afternoon, Laura finally arrived back at the hotel after a pleasant day of touring the historic steel town with Charlie and Pat. Of course, she couldn't miss seeing the huge iron statue of Vulcan, often referred to as the Moon over Birmingham because of his bare buns visible for all eyes. Known as the Roman god of fire and metal working, the gigantic man was cast for the 1904 World's Fair in St Louis and eventually placed at the top of Red Mountain to watch over the industrial city.

Laura was glad to finally catch up with Kevin. She wanted to tell him about last night, but she'd been sleeping soundly when he left early that morning. Laura hadn't expected the hostile greeting as she walked into the room.

"Where in the hell have you been?"

"What?" She was asking more about the tone of his voice than the actual question. Laura tossed her bright orange jacket on the bed and turned to face Kevin. "What are you talking about? I always spend the day out. You were at meetings all day." She tried to keep her voice even.

"Who were you with?" He snapped.

"Charlie and Pat. Why?"

"I heard you were with Adam."

"You heard wrong. What is your problem?"

"You were with him last night." Kevin hissed at her.

"Kevin, we were in the bar. I was with a hundred people, not just Adam."

"Didn't he walk you up here?"

"Yes, and we made mad passionate love on the floor while you were asleep on the bed. What in the hell are you getting at?" Now

Steel Illusions

Laura's tone was matching his octave for octave as she turned to face him.

"He was up here yesterday afternoon right after I left, wasn't he. Don't deny it."

"I'm not going to deny anything. He was here. We talked. Who told you?"

"Does it matter? Somebody saw him go into your room. He was a suspect, remember? You thought he might have engineered your kidnapping."

"I was wrong. He's the only one who'll take me seriously."

"So now that he's catering to you, he's a hero?"

"You're not being fair."

"Oh, I think I'm understanding things quite clearly. He met you in Vail, and now here."

"Kevin, you're jealous."

"Shouldn't I be?"

"No. Somebody's been following me, and Adam's the only one who's worried about it." Laura didn't hold anything back at this point. She was furious at Kevin's stupidity.

"Ridiculous! You're chasing shadows, and he's just feeding off your hysteria."

"Hysteria? You're calling me hysterical. Doesn't it bother you that you never found out who wanted to kill me?"

"Kill you? They kidnapped you. They weren't going to kill you."

"You have no idea what happened up there. You took care of everything here and assumed it was all over."

"You were scared. You only thought they were going to kill you."

"Kevin, don't be so naive. Not only did I suspect they wanted me dead, they told me. Didn't you ever wonder why I was able to give such a good description of Jake to the police. He didn't worry about keeping me blindfolded because he wasn't worried about my talking to anyone."

"Then why didn't he kill..." Kevin was interrupted by a knock at the door. "Who's there?" He snapped.

"It's me." Krista answered.

Kevin took the couple of steps necessary to open the door. He'd planned on telling her to meet them downstairs, but she rushed into the room with a huge man right behind her.

Krista was nearly frantic when she turned to face her mother. She'd completely missed the tension in the room. "He's out there, Mom."

"Jesus Christ, Laura. Now you've got our daughter afraid of your shadows too."

"Daddy, I'm serious."

"I've had enough. Pack your bags. We're leaving."

"Excuse me, Sir." David stepped up to give his side of the story.

"Who the hell are you?"

Krista was more horrified by her father's attitude than the episode with the man downstairs. "Dad!"

David rose to his full six foot four inch frame and moved to stand directly in front of Kevin. "My name is David Summerfield. And the man in the lobby is not a figment of your daughter's imagination or mine."

"What are you talking about?" Kevin ran his fingers through his hair as he walked over to the desk to take a drink of the Coke he'd brought up with him.

Krista offered the first part of the story. "When I came out of the lady's room, a man was standing just around the corner. I knew I had to be careful, but I was caught off guard. David was just a few steps away, and I was focused on him. The next thing I knew, there

Steel Illusions

was a gun in my ribs and a man telling me to be quiet, or he'd kill me."

"Part of this is my fault." David jumped in. "I'd picked up a paper and wasn't expecting Krista to be finished so quickly. When I looked up, all I saw was her walking away from me, arm in arm with a man. Her hair caught my attention. Not many people have the long curls."

"David yelled my name, and the man let me go. I'm not sure he even realized he had the wrong person until David yelled 'Krista'. I ran to David and the guy disappeared."

"Did you recognize him?" Kevin asked, finally with a sense of urgency.

"No. I'd know if I saw him again, but I don't know who he was."

"We told hotel security, but I don't think the man took us seriously." David added disappointedly.

"Imagine that." Krista looked directly at her father. "Ya know, Adam was right; we do look too much alike."

"Adam was right?" Kevin questioned.

"Yeah, Dad. Last night Mom was sure somebody was waiting for her when we left the bar. Adam made sure neither of us walked alone."

"And you..." Kevin looked at David, "...and you've become the appointed guardian of my daughter." There was no respect or gratitude in his voice.

Krista was appalled and embarrassed by her father's words. "David, I'm sorry you've been subjected to all of this. Let's go."

"Wait. Please wait." Kevin's tone was only slightly more civil. "I'm sorry." He held out a hand to David. "I owe you an apology."

David answered the handshake, but didn't say anything. He was trying not to pass judgement on the man.

"Thank you for saving Krista this afternoon. I didn't realize the danger."

"You're welcome."

"Krista, will you and David join us tonight at the banquet? We'll leave in the morning."

"I don't know. It's up to David. He certainly didn't come to babysit."

"I'm not letting you out of my sight. When I get on the plane back to Chicago, you're going with me." David's statement wasn't in a tone that left room for negotiation.

"Shit," Kevin said observing the electrical current between the two of them. "Another son-in-law."

"Daddy!" Krista was furious. "You have no right to say such a thing. You're more paranoid about your daughters getting married than Mom is about her shadows. At least her shadows are real." Krista walked out of the room with David close behind her.

"What's his problem?"

"I honestly don't know. He's usually pretty calm."

"He acted like he was jealous." David commented.

"Of what?"

"Me and Adam. He's been left on the outside of this."

"David, it's his own fault. He didn't want to believe us. I'm sorry about the son-in-law crack. My sister just got married last Saturday."

"How old is she?"

"Twenty-three."

"What's the big deal? She's certainly old enough."

"I think it had a lot to do with the fact that she'd only known Luke for three days."

"Three days? Hell, I'd probably want to kill the guy."

-343-

Steel Illusions

"I think he did, but Luke had just finished saving Mom and taken a bullet in the leg for his trouble."

"I see your point. It's hard to wring a guy's neck when you owe him a life."

Krista laughed. "Your colleagues wouldn't believe this."

"Neither would yours. You were right about your reputation."

"How do you know? You called the school. I should've guessed." She wasn't upset or even surprised.

"I hated not being able to place you at the hospital, so I called the Dean. He told me I was talking about the wrong girl."

"Told you no one would believe you."

"He gave me your description; that was all I needed."

"Oh, I'm sure it was flattering."

"You always wear your hair in a braid and wire rim glasses. I used that description along with the white lab coat and blue jeans; and, suddenly, there you were standing in front of me, questioning my medical opinion with the patient and other students listening intently. I didn't like you at all that day."

"Sorry, it's always been in my nature to ask questions.

"You should. I also found out you're at the top of your class Impressive."

"Anything else you want to know?"

"Not right now, but I'll ask if I need to."

"Thanks for being so understanding about my Dad. This whole thing has been hard for both of them to deal with."

"I'm sure. I'll tell him we'll wait for at least three weeks before getting married."

"Oh, I'm sure that'll make him feel better." Krista laughed heartily and relaxed.

"I'm serious, Krista."

R. Z. Crompton

She stopped walking but didn't look at him. This man was impossible. Then she looked up at him with a smile. "Ah, I get it. You're joking. You're trying to get even for the comment I made at the hospital that day when all the students were listening."

"Not at all. Marrying you will be getting even."

Krista was still looking up at him. She'd never been with a man who towered over her. This was ridiculous. "I know you're teasing me. You can't marry a student."

"You're wrong. I can't date a student. There's nothing in the University rules saying I can't marry a student."

"Why are we having this conversation?" Krista asked as they waited for the elevator. She couldn't deny her feelings about him had changed. But she certainly wasn't considering marrying anybody at this point in her life, not even Brad.

"I've never known anybody like you. There's so many sides to you. You're intense at school. Relaxed and funny out here, in spite of the danger. I've a funny feeling I'm still not seeing all of the sides you have to offer."

"So I'm a complicated woman. There's lots of us out there if you take the time to look."

"I have." They stepped into the elevator and the doors closed behind them. "What should I wear tonight. Is this a black tie event?"

"You're going?"

"Absolutely. Like I said, I'm not leaving until you get on the plane with me."

"A suit is fine. Ah, I wouldn't bring up the topic of marriage tonight."

"Is that your clinical answer for 'yes'?"

"No. I'm in school. Marriage and school don't mix." She couldn't help smiling up into his big brown eyes. The doors opened on the sixth floor. David took her arm as she stepped out. She appreciated the escort to her room. "Ya know, last night when Adam told me not to walk around the hotel alone, I really thought he was exaggerating."

"I'm glad he did, or we might not have had the chance to get to know each other." David looked around the corner of the hall. No one was there. "How long will it take you to change?"

"About an hour."

"I don't like the idea of leaving you down here alone. Do you want me to wait for you."

Having David in her room while she showered was more than she was ready for. "I won't open the door for anybody but you." She stated as she unlocked the door of her room.

David went in to look around. "Krista, it's only been an hour since a man tried to grab you. He knows you've seen him. I'll wait outside your door if you don't want me in your room."

"Don't be silly. I'm not going to leave you outside in the hall. It's just that..."

"What? You can't forget I'm the arrogant ass you have to go back to the University with."

"I wasn't going to say it like that. But yes. I'm finding it harder and harder to remember the roles we'll be going back to. Having you here..."

David walked up to her, but she didn't step back. He raised his hand to run his fingers through her hair. "Touching you makes it hard for me to think about original roles." She shook her head in agreement. "Then don't think, just enjoy. We don't simply enjoy often enough."

David lowered his head to kiss her on the forehead and down the side of her face before reaching her lips. Krista marveled at his tenderness. This was not the man she'd confronted in a hospital room at the University. Without noticing her own body's response, she was on her tiptoes reaching for the kiss. Then she was off the floor with her fingers running through his soft ebony hair.

"Put me down." She asked softly, hardly able to draw a breath. "I don't know if I'm supposed to feel safer with you in my room now or not."

David smiled at the disheveled look of her hair and the flushed cheeks. "I can tell you one thing for sure. We will not go back to Chicago and pretend this didn't happen."

"I'm really confused about my feelings and yours."

"So am I, and I've never felt so alive. Maybe confusion is good sometimes. Don't try to think it to death, Krista. You can analyze yourself out of anything. I know. Been there, done that. Now go shower. I'll feel as self conscious as you do when it's my turn to clean up."

She hadn't thought about it, but she was going to have the same opportunity to check out him and his room. "I'll hurry." Krista went through her drawer picking out what she needed. There was an inner smile as she pictured the dress she'd bought, *Daddy bought*, for the evening.

****

There was a pronounced chill in the air after David and Krista left the room. Kevin was chewing on a healthy serving of crow, and Laura was going to make sure he swallowed every piece. He deserved it. She took what she needed with her into the bathroom and closed the door. Kevin's first thought was to nurse his wounds at the bar, but realized he couldn't leave Laura alone. *She'd been right. Adam was right. Man, do I owe him an apology.* That gave him an idea.

Kevin picked-up the phone hoping Adam was in his room. "Adam?"

"Yeah?" Adam recognized the voice immediately. He'd drilled Kevin earlier for neglecting Laura. Needless to say, Kevin was pissed. Not only did he not believe there was anything to worry about, he felt Adam was over stepping his rights as a friend.

"I owe you. Krista was accosted by a man this afternoon. David stopped him. If you hadn't put them on guard last night, Krista might be dead."

"So, now you understand the danger."

"Right now, I'm trying to get my foot out of my mouth. Laura is taking a shower."

"At least you aren't leaving her alone."

"I could use a drink. I'll order. Would you come over and join me?"

"You gonna try to pop me in the face again?"

"No. I'm afraid Laura might throw something at me."

"You deserve it. Did you talk to her?"

"Not yet. I don't think she's ready to talk. I...we need to come up with some kind of plan for this evening. There's going to be a ton of people around tonight."

"Okay. Give me a few minutes to change."

Kevin picked up the phone again and ordered room service. A drink for him and Adam, along with a glass of champagne for Laura. He was going to have to grovel for a long time to make up for his accusations. He knocked on the bathroom door in an attempt to start the apology, but the sound of the shower was the only answer he got. *When she comes out and sees Adam here, that'll help smooth things over. At least, she'll know I'm ready to listen.*

Adam arrived before the drinks. He'd been relieved to get the call from Kevin but alarmed to find out about Krista. The mistake he'd been worried about was now reality. They'd have to be careful it didn't happen again.

"Thanks for coming. Laura's still in the bathroom."

The next knock on the door was the waiter with their drinks. Kevin motioned for the tray to be put on the table and signed the ticket. He handed Adam his glass and then took Laura's glass to the door. "Let's see how mad she still is." He winked at Adam. "My Dear, I ordered you something to drink." There was a pause before the door slowly opened, so Laura could reach out and take the glass Kevin held out for her.

"Thank you," she stated flatly.

"I guess there's hope." Kevin said when he looked back at Adam.

It took Laura another half hour before she'd cooled off enough to open the door. All she needed to do was put her dress on but wanted to wait until Kevin was in the shower. He didn't deserve the dress she'd bought to please him.

"Adam, I'm surprised to see you."

"He came to help us plan some kind of strategy for this evening."

"Strategy? You're finally going to admit there's a problem."

"Yes. I'm sorry I didn't listen to you." This wasn't the place for a full fledged 'suck-up'. "How was the champagne?"

"Fine. I'm ready for another one."

"Kevin, speaking of strategy, it'd be nice if we had some idea of who we're looking for."

"Not a single idea. Not one damn suggestion."

"Mason was clean when the police checked him out?"

"Nothing."

"Do either of you care about my opinion?"

"Sorry, if it's not Mason, who could it be?"

"The man who..."

Adam almost thought she was going to slip and give her secret away; but after a slight hesitation, she continued. "...who wanted to kill me got away. He knows I can identify him, and he knows I live in Tulsa."

"Jesus, I never even considered the possibility. Let's order another drink. I've got to think about this." Kevin made the call allowing ideas to sink in.

"Okay, let's assume it is him. How'd he follow you around without your seeing him?" Kevin asked.

"I didn't need to see him to know he was there."

"Why didn't he stay in the mountains where he was safe?"

"Let's ask him when we catch him." Laura knew he was playing devil's advocate, but it still seemed like a stupid question. "Maybe he wasn't finished with the task he was paid to do."

"How'd he know you were in Birmingham?"

"I don't know."

Adam had a very different picture in his mind than Kevin did. He saw Laura tied to a bed with a crazed mountain man on top of her. The man was crazy and that could be enough to bring him out of his sanctuary. "She's right, Kevin. He may be stupid enough to leave the mountains in order to finish his job."

The thought of Jake finishing his work made her skin crawl. If he took her again, there was no doubt in her mind that he'd finish his job. "That would explain the confusion with Krista. He never saw her in the mountains." She was suddenly very worried about her daughter. "Have you talked to Krista. I don't like the idea of her being alone."

"I'll ring her room." Kevin offered picking up the receiver.

The answering voice was deep, too deep. Kevin new who it belonged to. "Is Krista there?" The frown on Kevin's face caught Laura's attention. There was silence for several seconds before Kevin snapped, "What in the hell is he doing in your room while you shower?"

"Stop acting so innocent, Daddy. He's standing guard. What did you think he was doing?"

"Damn it, Krista. He's a man."

"Ya think so? I hadn't noticed." She teased.

"Don't get smart."

"I'm twenty-five. Having a man in my room is my choice. You were calling, I assume, to see if I was fine. Yes, my appointed guardian is here. How's Mom?"

"Better. She's talking to me."

"That's good. What time do you want us to meet you?"

"Half hour."

"Fine. Just so you know, we are going to David's room, 986, so he can get ready. Do you want to come by?"

"You're being smart with me."

"Bye, Daddy." He's acting more like himself she announced to the man sitting on the sofa watching the news. "I just have to get my dress on." Krista took the black dress off the hanger and into the bathroom. She wiggled into the tight lycra, added the dangling earrings and high heels and opened the door.

David looked up when he heard the door open. "Holy cow!" His first reaction was pleasing. "You're gorgeous."

"Thank you. Would you zip me, please?"

"Up or down?" David asked as he walked up behind her. The perfume was inebriating. "If you marry me tonight, your dad won't be mad at your sister anymore."

"No, but he'll certainly want to kill you."

His hands left a burning sensation on her skin as he moved the zipper to the top of the dress. He touched her neck and then moved out to her shoulders and down her arms. The black mesh offered no boundary between his hands and her skin. It was all she could do to move away from him.

"Maybe I should wait with Mom while you get ready."
"I'll be good. I promise."
"That's what I'm afraid of."
"Krista, I'm surprised you'd even think such a thing. Let's go." David opened the door and peered down the hall before letting her walk out. His arm was possessively wrapped around her waist. Nothing was going to stop him from claiming her as his own, except maybe her. *Don't push too fast. This isn't the operating room, well not the hospital operating room.* The elevator doors opened and closed before the man stepped out of the shadows.

CHAPTER TWENTY-TWO

Kevin saw his daughter standing next to the well-dressed man who'd answered the phone. They made a fine looking couple, and he definitely had the look of a possessed man. Krista turned slightly allowing the light to filter through her skirt.

"I suppose my money paid for that dress?" He asked Laura.

"She looks stunning doesn't she?"

"When did I miss the fact that both my babies were all grown up? I miss them."

"So do I." Laura had a tight grip on Kevin's arm.

"Do you think she's in love with the man?"

"Yeah, I do."

"What about Brad?"

"He was a very good friend. David will be her hero."

The ballroom was packed with people when they arrived. Adam came up with a drink for each of them. "I haven't seen anybody who looks suspicious. Laura, you're going to have to watch everybody. You're the only one who's seen this guy; and after his grab for Krista, I don't think he's going to waste anymore time."

"I agree." Kevin scanned the crowded room.

"Kevin, I had another idea. What if it's somebody registered for the meeting whom we just never considered?"

"I guess it's possible."

"Did you ever check the registration to see how many new members are here?"

"I never thought about it. The registration room is just down the hall. I can go look. It's easy to pull out the first-timers."

"I'll go with you. We can look faster together."

"Laura, we're going to the office for a minute. Do you want to come?"

"No, I'll join Krista and David."

Laura examined the beauty of the room. Kevin had told her about the special wallpaper airbrushed by students from Auburn University. Peach and various shades of green complimented each other throughout the room. Within minutes Laura was caught up in the excitement of mingling through the crowd. Black tended to be the color of the evening with most of the ladies decked out in various lengths and styles.

"Hello, you two. You both look very handsome this evening."

"Thanks, Mom. Where's Dad?"

"He went to the office. Why?"

"So he finally believes us?"

"I guess so. I hope we can put a stop to this. I don't know how much longer I can play the game. Thanks, David for looking after Krista. I do appreciate your offer to get her back to the University. I won't feel safe until she's out of here."

"No problem."

"Krista, if I'd known you'd be in danger, I wouldn't have let you come."

"Stop worrying. We had no idea this guy would mistake me for you. I'm fine."

"When are you going to fly back?" Laura looked at David.

"It's up to Krista. I have more flexibility than she does."

Krista was glad he didn't mention any of their more intimate conversation. This wasn't the time to add more to the family stress level. "My flight out is for tomorrow afternoon. I can wait if you need me."

"Please go. Your father and I will worry less if we know you're safe."

Krista looked up into David's face. She wondered how safe her father thought she'd be leaving with a man who'd joked about marrying her. "We'll see what happens."

David looked down into the light blue eyes and winked at her. He knew why she'd not jumped on leaving right away. They still had to come up with a definition for their new relationship; and at this point he didn't think he'd be able to convince her to marry him before going back. *Man, would that shock everybody on campus. I'd love it almost as much as I think I love her. There's always tonight, and she can't be left alone.* His arm came in tight around her, and Laura didn't miss a thing.

An exchange which can pass only between a mother and daughter went from Laura to Krista. The smile on Krista's face was the only answer Laura needed. "I'm not going to tell your father. That's your job."

"What?" David missed the exchange. "Tell him what?"

"Mom, you could tell him when you get back home."

"I'll tell him myself." David announced. "Besides, he might as well know the whole story before we leave."

"He might shoot you, too."

"What's he going to say after we spend the night together?"

"Wo! Wait a minute! Where'd you come up with that?"

"You can't stay alone. Since I'm your appointed guardian, as your father put it, then I'm the one who should stay with you. You can't sleep in the same bed as your folks, can you?"

Krista hadn't gotten that far in the planning. "David, you can't possibly be considering asking my father for permission to sleep with me."

"Well, now that you bring it up, it was on my mind."

"David!" Krista flushed from the embarrassment of the conversation. "This isn't something I talk about with my parents."

Steel Illusions

"I'm enjoying it, Honey. Go ahead." Laura was extracting extreme pleasure from this. "The entertainment is relieving my stress."

"Mom, don't encourage him. Really, David, you're as bad as my father. You two should get along well together if he doesn't shoot you first."

"Is that a 'yes'?"

"For which question?"

"Marry me?"

"Wow! He is serious." Laura smirked at her daughter.

"Is this what happened to Megan?"

"No. They had a fight first. Ya know, Brad predicted this. I didn't believe him at first. But he was right."

"Who's Brad."

"An old friend." Laura threw out before Krista could stick her foot in her mouth.

"So?"

Krista hesitated. Why so quick? Nothing in her life had ever been done without analyzing and re-analyzing. This was most important decision in her life and she was considering it on a whim.

David didn't want to hear "no", so he offered his thoughts. "I know what you're thinking. You're just like me. It's too fast. I don't know him. We should wait."

In a crowd of five hundred people, this man was working at a proposal he never thought he'd make to anyone. "I'd given up hope of ever getting married. As you know, my reputation at the University makes me less than the ideal catch for women like you."

He was right about that. At school, she could never see herself sitting at the same table with the man, and she'd never consider going on a date with him. "I must be absolutely crazy to even listen to this."

"Tell me, doesn't it feel good?"

R. Z. Crompton

She couldn't deny her excitement when he touched her or even looked at her for that matter. "I'll tell you in the morning." She'd hoped that answer would give her the time she needed to rationalize her own questions. But he was right about one thing, he did feel good. She'd never felt this way with Brad. David took her breath away. He was the storybook hero she'd always dreamed about. Finding the perfect mate was a stupid idea. There was no such thing as perfect. However, finding an exciting mate and one she could build a life with was what she wanted. She really wanted to tell him "yes." Now she understood what the guys were going through. Her body and her brain were having a hard time coming up with the same answer.

"Krista, I wouldn't mention this to your father. He certainly won't condone David's style of 'guardianship'."

Krista blushed again. The nightly arrangements hadn't been considered in her answer. "You're right. Why has he been gone so long?"

"Do you want us to go look for him?" David asked.

"Please. I didn't expect them to be gone so long. I'll stay here. Nothing can happen with all these people around."

David was glad to get Krista out of the crowded room. He wasn't used to spilling his emotions in public. He knew her logic was getting in the way.

"Laura, may I get you a drink?"

Laura was surprised to hear some one calling her name so quickly after Krista and David had walked away. "Anthony, I didn't know you were here. You don't usually come to the Globetrotters."

"This is my first one. Where's Kevin? I was going to talk to him, but I can't find him."

"He's checking on some records. I'll go to the bar with you." Laura walked in front of Anthony. Faces from the week before automatically flashed through her mind. She tried to stop the

Steel Illusions

flashback, but her mind worked in spite of her attempts to focus on a conversation. "So have you liked your first conference?" Laura turned as the face of Mac flashed into her mind. The man behind her looked strangely like the man who'd stalked her at Cassidy's. She felt her hair stand on end. Without making a scene, she tried to look through the crowd for Kevin or Adam. *Where in the hell are you guys? It's Anthony.* She wanted to scream out to everybody, but there was still no proof: only shadows, only images in her mind.

When Laura turned around again, Anthony was gone. Laura picked-up her gin and tonic with a sigh of relief. For some reason Anthony sparked her memory. He had something to do with Mac, but she couldn't put it together. Even when Kevin came back into the room, what would she say? Anthony Mason's here. He had every right to be here.

Finding Kevin and Adam was all Laura could think of. What if something had happened to them? What if she'd sent Krista and David to the same fate? Laura calculated that the office they'd gone to was right out the side door of the ballroom. She looked around the room again; but seeing no allies, she put her glass down and headed for the door.

To Laura's surprise the door opened to a long, dimly lite gray hallway rather than the carpeted lobby she'd expected. *The door has to be here some where.* The first set of doors she tried were locked. Laura turned the corner and tried again, but this door too was locked.

Laura thought it would be wiser to turn around at this point and go back the way she'd come. It'd be too easy to get lost back here. Laura opened the door to Anthony Mason. He pushed her back into the hallway before she had the chance to react. Her purse caught in the door leaving it slightly ajar as he forced her up against the wall. The small gun he held in his hand answered an important question for her; he was the bad guy.

"You were Waterman's partner?"

"I was the brains of the partnership, he was the arrogant ass. He left too many loose ends hanging out there. That was his mistake. I had only two: you and Mac."

"You set up the kidnapping? And the avalanche in the Pass? Why?"

"No, I didn't do the avalanche. You give me too much credit. Jake did that with a timer. We didn't want the rescuers running off to find you. Jake needed time to get you hidden away. He was stupid to not keep better watch."

"Why the kidnapping?"

"Waterman wanted to watch Kevin fall apart. I told him he should just kill him outright; but no, he had to play his fuckin' games."

"But why now, the police cleared you. There was absolutely nothing to tie you to all of this."

"Nothing but you. I saw the reaction tonight when I walked up to you. The beginning of recognition. You'd have put me and Mac together eventually."

In the dimness, she knew why Anthony had sparked the memories she'd tried to bury. Mac was his brother. She met the man only once, but Kevin would know him. Anthony didn't have the beard or the black clothes, but Laura had the connection. It didn't matter now what memories he sparked; he was going to kill her. *I have to do something.* Anthony pushed himself up against her body with the gun in between them.

"Tell me, Laura, was it good in the mountains?" He slurred to her in a raspy voice.

"What?" Laura was sure she misunderstood the intent of his question. How could he possibly know what happened up there.

Using his free hand, he ran his fingers around her hips. "You know what I mean. Jake had special plans for you."

Steel Illusions

Laura struggled to get away, but his body and the wall were like a vice grip crushing her ribs. "How do you know?"

"He's quite a colorful character. Known him for years. We have similar taste in entertainment. I like to use ropes too."

"Jake's an animal; and you're no better."

"Oh, yes. I'm better and smarter. Now move, we're going to find us a fun place further away from this door."

Laura had no time left. If she was going to make a move it had to be now. With all her force she drove her knee into his groin. Anthony groaned as he doubled over. "You bitch."

Laura bolted down the concrete hallway turning left into another long corridor. She dashed around the tables and toward the doors at the end. It was her last chance to get away. She could hear the man behind her trying the same doors she'd checked just seconds earlier.

Laura burst through the doors to what she considered the loading dock. Concrete was everywhere. Garage doors lined the area which was void of any other humans. Laura, glad she had not worn a tight skirt, ran up the long drive looking for a way out. She heard Anthony enter the docks behind her.

Anthony wasn't going to let her get away again. The black skirt flashed in front of him. There was nowhere for her to go. "Laura, you can't get away. You should've died up on that mountain. I paid Jake a hell of a lot of money to take care of you and Mac. I couldn't have either one of you coming back to talk. I'm going to have a long talk with Jake for screwing this up." He could hear her heavy breathing, and then she burst out from behind a concrete slab just in front of him. She was running scared; he had to grab her before she escaped the long drive.

If there's a guardian angel out there, now would be a good time to show up. She searched for something, anything to help. A light.

Thank you. Laura saw a light coming from the crack between two doors. Her lungs burned. Her body glistened with perspiration causing the shear material to stick to her legs as she ran past the concrete bunkers. Would the door be locked? If it was locked, she'd be at his mercy; and she didn't think mercy was a strong part of his character.

****

"Adam, the door is locked."

"What. Can't be. Beat on it. I'll check the other exits."

"Hey! Help!" Kevin hit the door first with his fist and then with a chair. "Any luck?"

"Nothing, we're locked in. Ya think this was an accident?"

"Not a chance in hell. I sure hope Laura stayed with Krista and David. That guy's so damn big, he'll scare anybody off." Kevin hit on the door again with the heavy walnut chair. He could barely pick it up let alone swing it. "I still want to know how Krista met the man and what he's doing here. I don't think I like him. He's so sure of himself."

This time Adam rammed the chair into the locked door. "He's one of her instructors."

"Another doctor? Just what I need; more attitude." Kevin took his turn, but the door didn't budge.

"At least, he's not a bum. Help. Help." Adam yelled at the door.

"Dad, is that you?"

"Krista get us out of here. Where's David?"

"Right here."

"Your mother?"

"In the ballroom. Why?"

"It's a set-up. Hurry, get us out..."

Steel Illusions

Kevin nearly fell out of the room as David pulled the door open. "How'd you do that? You lift weights, too?"

David held up the pipe in his hand. "No, Sir. This was wedged in the doors."

"C'mon. We've got to find your mom."

"Did you find anything in the records?" Krista asked as they hurried for the ballroom.

"Yeah, Mason. He's never attended the conference before. So why now?"

"Dad, I thought he was cleared by the police?"

"So we were wrong again. What's new. We haven't guessed anything right yet. It's a good thing I'm not a cop. I'd never put all the pieces together."

Adam threw out the next question. "Why'd he register at all? He could've merged in with the rest of us, and we wouldn't have known any better.

"I would've. If I had seen him or heard he was here, I would have known he wasn't registered."

"Why's he after Mom?"

"I don't know, and I'm not guessing anymore." They entered the large room, searching. But no Laura. "You two go that way. Ask people if they've seen her. Adam, through the middle. Hey, ask about Mason too."

The four searchers scattered to look for Laura. There was nothing. Krista didn't like playing the search game again. Life could be cut short with no warning. Her mother had been threatened twice in one week. The fear she'd felt when she'd been sucked up by the snow overwhelmed her, causing her stride to falter. For just an instant all these people were waves of snowing stopping her from reaching her goal. Beads of sweat formed on her brow, and she stopped abruptly.

"What's wrong?" David asked.

"The crowd is like the avalanche; hunks of snow beating me down. I couldn't breath."

"Are you okay?"

"Yes, I just had the strangest 'reality check'. Life can end so easily." She knew what her answer to David's question would be.

"Let me go first." David was a bulldozer clearing the way for her. But there was no sign of her mother.

Adam reached the other side of the room first and turned to continue his visual search while he waited for Kevin. Laura didn't just disappear. *Think! Think. Why'd she leave this room. She was safe here."*

Kevin looked frantic when he came up to Adam with no Laura by his side. "People don't just disappear."

"Yeah, Dad. They do. This is how it was for us in Vail, crowd and everything. We came back; she was gone." Krista offered when she came up behind David.

"I don't believe this. There's over five hundred people in this room, and most of them know her. How can she vanish without one of them seeing her."

"I'll get security." Adam rushed away.

"Dad, if so many of them know Mom, why don't you get them to help look for her. There's a microphone on the podium."

"Good idea. We can cover the hotel a lot faster if we have some help." Kevin raced through the crowd and turned on the microphone. "May I have your attention please? Please?" The second request displayed the emotion rushing through his body. "Laura is missing. We suspect she was abducted or running from Anthony Mason. I need your help to find her. Please search the hallways, restrooms, closets and any other place you come across. Mason is dangerous. If you find them, pretend the party is simply spreading out. Come back here when you find something or when you're finished."

Steel Illusions

What fun! Everybody moved out quickly enjoying the new part of the program. The crowd broke into smaller groups; some going up the escalators, and some to the mall area which was almost ready to close down. Other's covered the entire lobby. Kevin was grateful for the excited participation. When the ballroom was empty, Kevin saw the purse caught in the door. He knew it was Laura's even before he picked it up. Krista watched her father reach down for the bag. Now he understood how frustrating she and Megan had felt.

"We'll find her, Dad." She put her arms around her father trying to give him the reassurance David had given her.

"Kevin, where is everybody?" Adam walked in with the security staff behind him.

"They all went to search for Laura."

"All of them? That's a hell of a big search team."

"I didn't know what else to do. I found this by the exit door."

"Hers? Kevin, why would she leave this room and put herself in danger. She was safe here."

"He either confronted her, or she went to find us." Kevin looked at the door. "Do you think she might have thought this led to the room we were in? We were just around the corner. What's down here?" Kevin asked the security guard.

"It leads to the loading dock if she makes it all the way to the end. Otherwise it's a series of locked doors."

"Let's check it out. Krista, you and David wait here with a couple of the guards. You come with us." Kevin looked at the man standing closest to him.

Quickly they moved down the dreary concrete hallway. They made the left turn with the only sound being the beat of the shoes as they hurried through the maze of tables and carts left by the waiters. Adam and Kevin checked the locked doors and hurried on. There was nothing.

R. Z. Crompton

****

Laura ran for the door. She couldn't see a lock breaking the light streaming through the slit between the doors. It was her last hope of escape. She threw herself at the door; and, to her surprise, it flew open. Anthony cursed behind her. Bright lights encouraged her on, and the concrete floor gave way to white and black tile. Another second and Laura burst through the mall entrance at the lower level. *Thank you, God.* Several people noticed the abrupt entrance she made into mall. Not many women appeared out of nowhere dressed in suede and black chiffon, and the eyes followed her as she made the hasty search for the corridor leading back to the hotel. The man following a few steps behind went almost unnoticed.

Another set of eyes caught the movement on the lower level. "Look, I see her." Elaine pulled Bob after her.

"Mason's right behind her. We gotta help her. It's to far to get back to the hotel for help."

"What are we going to do?"

"I'll figure it out when we get closer. She's got to get up the escalator."

"Follow the railing. Maybe we can get her attention." Elaine and Bob moved quickly watching Laura's progress toward the escalators.

"What if I throw my shoe over the rail?"

"That's too big. He'll see it. She has to go up in order to reach the hotel. Elaine, go over by the wall. If she starts to go the wrong way, I'll wave; you let out one of your wolf call whistles."

Bob watched Laura move quickly toward the center of the mall. He could tell by her puffing she'd been running hard. She couldn't out run the guy much longer. At the moment of hesitation, he waved at

Steel Illusions

Elaine. The loud whistle caught Laura's attention. She saw the slight wave just above the rail, and she moved in the direction of the escalator.

Laura had almost missed the moving steps, but somebody knew she was in trouble. *Thank you, God.* Anthony was only a store length away from her. He hadn't been able to run full out. It would've attracted too much attention. She was afraid he'd be able to make up the distance by taking the escalator steps two or three at a time. She was at the steps, but her heels hindered her ability to take more than one at a time. Laura felt him reach out for her. The skirt of her dress tore at the waist line.

"Ahhhh." Laura screamed.

He reached again for her grasping her ankle.

Laura lunged forward grabbing the moving rail, and she pulled herself out of his grip. Next time he'd have her. First one step then the other, she was out of the shoes and leaping the steps.

"How are we going to stop him?" Elaine whispered.

"I'll show you. Stand back from the rail now. Let her get past us."

Laura took her last step off the escalator running for the hotel entrance. She didn't see her heroes behind her step in front of the pursuer. Locked in a tight lover's embrace, Elaine and Bob stepped in between Laura and Mason. He stumbled over them landing all three on the floor.

"Hey, Anthony. Sorry we got in the way. We weren't paying attention." Bob said as he pulled Elaine up beside him.

"No, shit!" He hissed, desperate and out of time. Anthony tried to push past them when he saw Laura stop and turn around, but Elaine took her turn to apologize.

"We're really sorry. You okay?"

"Yeah, I'm fine."

Bob and Elaine stood staunchly in front of the moving stairs, not letting Mason pass them. Elaine feigned concern about his well being, but didn't budge when he tried to move around her.

"Look, there's Kevin." Elaine yelled.

When Mason turned to look, his foot caught the top of the last stair knocking him off balance. A little shuffle of the feet from Bob and his partner sent Mason stumbling down the stairs. Twisting his body around he came up, without his gun, at Kevin's feet.

"You son-of-a-bitch," Kevin hissed. He grabbed Mason by the collar and smashed him in the face with his fist.

Mason staggered back a step and then threw his own punch at Kevin's jaw. Kevin was surprised at the lack of pain. Adrenaline surged through him blocking the pain. This was the man who'd threatened his wife. All frustration of the last ten days fueled punch after punch. This man wouldn't be hurting Laura ever again.

Laura watched with Bob and Elaine on each side of her as Kevin exploded, beating Mason into unconsciousness. Adam was still at the bottom of the escalator, and he watched while Kevin exercised his wrath on the man who'd tried to destroy him and his family. The police were already coming from the direction of the hotel. They'd stop Kevin from killing the man.

"Elaine, Bob, thank you so much." Laura with tears of relief streaming down her face, threw her arms around the two friends.

"Are you all right?"

"Only thanks to you."

"Nothing to it. We were glad to help. Exciting addition to the normal program."

Laura tried to laugh, but was still running short on humor. "Not one I want to replay if you don't mind."

A crowd was starting to form at the top of the steps. Anyone who'd seen the police enter the lobby followed them out to the excitement. They watched the police take charge of the bloody Mason and then applauded Kevin and Adam as they climbed the escalator two steps at a time. Kevin drew Laura into his arms and whispered, "I'll never doubt your intuition again." He kissed her deeply. "This time, this time it's over, My Dear."

This time Laura knew he was right. Jake would still be in her shadows; but, hopefully, he'd stay deep in the shadows of the mountains and her mind, away from her daily reality.

"Adam, thank you. Everybody, thank you for helping us."

Cheers were simultaneous. "Isn't it time to party?" someone yelled.

Again the cheers, and the crowd headed for the lounge. Krista and David rushed down the long lobby to meet Laura and Kevin. Kevin noticed the familiar touch of the hand on his daughter's waist. Considering the fact that the danger was over, David was certainly sticking close to his daughter. *Shit, another son-in-law. Every time Laura needs a search party, I lose a daughter. Good thing I've only got two daughters.*

Drinks were ordered by everyone as Laura was asked to tell the story over and over about how she was caught in the maze behind the ballroom, and how Bob and Elaine saved her with a big wolf whistle and faking a lovers' passionate kiss.

"I knew it was you, Elaine." Laura claimed. "Nobody whistles the way you do."

"You did not. You were too busy trying to get your shoes off. And who said we were pretending?"

Kevin didn't really participate in the conversation. He was still stunned by what happened. If Adam hadn't forced him to pay attention, he might have lost her forever. Adam was sitting several places away, so Kevin couldn't talk to him directly.

"Excuse me, Everyone." He waited for the din to settle. "I owe a great debt of gratitude to Adam." He raised his glass to toast his friend. "I'm sorry I doubted you. Thank you for making me pay attention to what's important." Adam raised his glass to acknowledge his acceptance of the apology.

Kevin looked around again, but didn't see his daughter. "Laura, do you know where Krista and David are? I wanted to thank him for his help."

"I think they might want to be alone."

"Why?"

"Kevin, think about."

"I can't. I don't even know who this guy is. He doesn't need to protect her anymore." Kevin got up to leave in spite of Laura's protest. He walked into the lobby and saw his daughter in the arms of the very tall and disgustingly handsome young man. He was the type of guy who could break a father's heart when he stole his daughter away.

"Krista, there you are. I wanted to thank David for his help this weekend."

"It was my pleasure, Sir."

"When will you be going back to Chicago?"

David had the feeling the question was searching for more than the obvious "tomorrow". "We are going back on Krista's original flight. Mine was supposed to go earlier, but I've already rescheduled it."

"I see."

David looked at Krista wondering if she had the courage to tell her father the truth. He knew she had reservations about dropping another wedding in his lap, but he'd promised to let her do the talking.

"Dad, David and I are going to ...."

"Yes, going to what?"

"Well, you know how things can just happen."

Steel Illusions

"Yes, I'd say tonight is a good example of how things can 'just' happen, so what are you saying?" Kevin winked at David. He was going to make her struggle to tell him.

"Well, we decided that we're old enough to..."

"To what? Sleep together?"

"Dad! I can't believe you'd say that." Krista was blushing. "You and Mother deserve each other. You're both terrible."

"If you're having so much trouble telling me what's on your mind, then maybe he's not worth it."

"Daddy! We want to get married before we go back to school."

"I know."

"You know! Why didn't you say so?"

"It was more fun to watch you play around the words. You may be extremely intelligent, but I can still win the word games." Kevin held his hand out to David. "It's good to have you as part of the family. If you're willing to take her even after this weekend, you're a good man."

"Thank you, Sir."

"If you hurt her, I'll skin you alive."

"I'm sure you will."

"When and where do you plan for this to take place?"

"In a few minutes, Dad. A minister is on his way down. I figured we'd have plenty of witnesses if we did it here."

Within minutes, the Globetrotters were gathered in the lobby of the elegant Wynfrey Hotel to watch the wedding of Krista and David.

"So, Laura, what are we going to do for excitement at the next conference?" Elaine asked.

"I hope nothing." Everybody laughed and waited for Kevin to give his daughter away.

R. Z. Crompton

White Gold
Watch for the sequel to Steel Illusions. Luke and Megan will face the rigors of the wilderness when shadows from Luke's past try to destroy them.

Death By Association
Crompton's first novel is available from the Zoller Publishing, Inc., P.O. Box 461661, Aurora, CO 80046.